Proactive Intelligence

John J. McGonagle · Carolyn M. Vella

Proactive Intelligence

The Successful Executive's Guide to Intelligence

 Springer

John J. McGonagle
The Helicon Group
Faith Drive 276
Blandon
PA 19510, USA

Carolyn M. Vella
The Helicon Group
Faith Drive 276
Blandon
PA 19510, USA

ISBN 978-1-4471-5911-7   ISBN 978-1-4471-2742-0 (eBook)
DOI 10.1007/978-1-4471-2742-0
Springer London Heidelberg New York Dordrecht

British Library Cataloguing in Publication Data
A catalogue record for this book is available from the British Library

Cover design: eStudio Calamar S.L.

Printed on acid-free paper

Springer is part of Springer Science+Business Media (www.springer.com)

Preface

We have discussed some of the concepts in this book before in many of the books (McGonagle and Vella 1987, 1988, 1990, 1996, 1998, 1999, 2002, 2003) and articles we have written on competitive intelligence (CI). However, this book marks a new direction—it focuses on competitive intelligence collected by, analyzed by, and used by—**you alone**. This is not going to be a guide on how to work full-time as a CI specialist. Instead, we want you to learn enough about CI, and what you can do with it, to enable you to do your job, whatever that is, better.

We are not going to go in-depth into the history of CI. However, we think it is useful for you to look at its roots, to see what lessons those who do CI full-time originally learned. This will help you have a firmer foundation in CI's basics.

One of the forces that helped to develop competitive intelligence was its link to corporate strategy first developed in the seminal strategy books by Professor Michael Porter (1980, 1985, 1990). In Porter's vision of a competitive strategy achieving a competitive advantage, competitive intelligence is the link that enables a firm to develop its competitive strategy, which, when executed will produce a competitive advantage for it.

Over time, CI spread its mantle. CI professionals began providing intelligence not only on competitors' strategy, but also on their tactics; not just on macro-level issues such as long-term investments, but on micro-level issues, such as pricing and product positioning. Regardless of the targets, it has been the goal of the CI professional to provide intelligence that the end-user will, well, *use*. As the end-user is not the same as the intelligence analyst and collector, CI historically ran into a fundamental disconnect. That is, while CI, very good CI, can be provided to the decision-maker, there is nothing requiring the end-user, the decision-maker, to use it to make a decision.

A second force that has shaped CI was environmental scanning. There, the goal was for business decision-makers to review their entire operating environment; political, economic, cultural and social, as well as competitive. Since decision-makers are in the business of making decisions, and not of constantly watching, environmental scanning quickly became something which was handed off to others to do. It was seen as serving as an early warning system to decision-makers of trends, facts, or events that could adversely or beneficially impact a business.

Environmental scanning is still with us, but it is not as robust as classic CI. Keep in mind the analogy developed by Herb Meyer, a pioneer in CI, which carries with it the implication environmental scanning is a part of competitive intelligence:

> Like radar, a Business Intelligence System does not tell the executive—the pilot as it were—what to do. It merely illuminates what is going on out there on the assumption that with good information a competent executive will nearly always respond appropriately (Meyer 1991, p. xi)

A third force that helped to develop classical CI was its use by technology teams in a wide variety of industries to develop and use access to data on pending research and development. The goal was to help develop an understanding of where their competitors were, are, and will go. Technology-oriented intelligence grew out of this environment. It still enjoys the creativity, cross-fertilization, and intellectual stimulation that having CI professionals work directly with the decision-makers generates.

So, what does all this mean for you? It means that the tools and techniques that will enable you to produce your own CI for your consumption are out there, and have been honed by decades of work. But, you cannot just adopt them—you have to adapt them.

Why? Because, when you get finished reading this book, you will still be the data collector, the analyst, and the end-user. But traditional CI is premised on a reactive, two-part relationship—that is, a CI professional responding to what an end-user identifies as a need, usually the result of seeing another new threat. But, by doing this yourself, you can turn CI from being reactive to being proactive. *As the decision-maker, you can get what CI you need, when you need it, and then use it almost seamlessly.*

Why not just always use an internal CI team or an outside CI consultant? There is nothing wrong with those two options, and we will discuss them later. However, just as managers and executives must learn to understand and use accounting controls, they should now learn to integrate CI into their own work processes. This does not mean that your firm cannot effectively use a separate CI unit; it still needs access to accountants too. It means CI can help you do your job—better. And, if you have an internal CI unit or use an outside CI consultant, doing some CI yourself first will make you into a better client for them. It will also make them into better intelligence providers for you.

To help you understand how powerful CI can be we have included a number of case studies as examples throughout this book. If we did not footnote it, it is something from our own professional experience. That means we still have to protect the client's confidentiality. Any others are documented, coming from books, articles, presentations, and even web sites.

Throughout the book, you may see us repeating a point or two. No, that is not a mistake. What we are doing is writing a book for you to use. At first, you may read this from start to finish, or then again, you may not. In any case, we expect that you will dip into the book again. So, we have tried to leave important points or tips where you will need them, when you need them. Think of this book as a guide to self-education, not as a GPS car navigation system, telling you exactly what to do and exactly when to do it.

References

McGonagle JJ, Vella CM (1990) Outsmarting the competition: practical approaches to finding and using competitive information. Sourcebooks, Naperville

McGonagle JJ, Vella CM (1996) A new archetype for competitive intelligence. Quorum Books, Westport

McGonagle JJ, Vella CM (1998) Protecting your company against competitive intelligence. Quorum Books, Westport

McGonagle JJ, Vella CM (1999) The internet age of competitive intelligence. Quorum Books, Westport

McGonagle JJ, Vella CM (2002) Bottom line competitive intelligence. Quorum Books, West-port

McGonagle JJ, Vella CM (2003) The manager's guide to competitive intelligence. Praeger Publishers, Westport

Meyer HE (1991) Real-World intelligence—New Edition. Grove Weidenfeld, New York

Porter ME (1980) Competitive strategy. Free Press, New York

Porter ME (1985) Competitive advantage. Free Press, New York

Porter ME (1990) The competitive advantage of nations. Free Press, New York

Vella CM, McGonagle JJ (1987) Competitive intelligence in the computer age. Quorum Books, Westport

Vella CM, McGonagle JJ (1988) Improved business planning using competitive intelligence. Quorum Books, Westport

Blandon, October 2011   John J. McGonagle
                                                                         Carolyn M. Vella

Acknowledgments

The authors would like to thank the following for allowing us to advance and then refine some of the ideas and concepts over the past years which we bring to you now:

The Editors and Publishers of the following publications:

- CIO.com—Business Technology Leadership
- Competitive Intelligence Magazine
- Competitive Intelligence Review
- Engineering Management Review
- IAFIE NEWS
- Journal of Competitive Intelligence and Management
- Legal Management
- MRA's Alert Magazine
- SCIP.insight
- SCIP.Online
- Security Management
- The [Newark] Star Ledger
- The Information Management Journal
- The Sunday Times
- Vault.com's Case Closed: The Career Newsletter for Consulting

Meeting and seminar sponsors with the following:

- Aligning Medical Affairs & Services for Success Conference
- American Chemical Society
- American Society for Industrial Security
- ARMA International
- Association of Independent Information Professionals
- Association of Strategic Planning
- Business Threat Awareness Council
- Construction Market Research Council
- Defense Industry Initiative

- High Technology Learning Corporation
- Institute for Competitive Intelligence
- Professional Pricing Society
- Society of Competitive Intelligence Professionals and its Atlanta, D. C., Kansas City, Mercyhurst College, New Jersey, New York, Philadelphia, Richmond, and St. Louis Chapters
- Strategic & Competitive Intelligence Professionals—New Jersey Chapter
- Society of Manufacturing Executives
- Special Libraries Association
- SUNY New Paltz Business School
- University of Central Missouri

Also, we want to extend a special thanks to Sanford J. Piltch, Esq. of Allentown, PA, for reviewing our discussion on intellectual property protection.

Contents

Chapter 1
Competitive Intelligence Lingo

Throughout this book, we have tried to avoid using too many Competitive Intelligence (CI)-specific terms—like CI. However, it is impossible to avoid that. In our quest to make this book as user-friendly as possible, we will list here some of the terms of art that we have used, as well as some that you might run into later on when digging deeper into CI.

For those readers who do not think they need to read this before getting right into the book, we suggest bookmarking this chapter for future user.

We decided to put the lingo up front rather than in a glossary because this is a learning experience for you. And using a glossary is more appropriate for a reference book—particularly since most readers just skip it anyway.

Analysis: In general, an examination of facts and data to provide a basis for effective decisions. Analysis often involves the determination of cause-and-effect relationships. In CI, analysis involves evaluating and interpreting the facts and raw data to provide finished intelligence to support effective decision-making.

Anomaly: An incongruity or inconsistency.

Baldrige Performance Excellence Program: Educates for profit and non-profit organizations in performance excellence management and administers the annual Malcolm Baldrige National Quality Award. Formerly known as the Malcolm Baldrige National Quality Program.

Barriers to Entry/Exit: Obstacles that make it difficult to enter/leave a particular market.

BCG matrix: Boston Consulting Group matrix.

Benchmarking: Analyzing what you do, quantifying it, and then finding ways that other firms do it better, or better and differently (if at all). Then, you adapt (not simply adopt) what you have learned to your own firm. Organizations engage in benchmarking activities to understand the current dimensions of world-class performance and to achieve discontinuous (non-incremental) or breakthrough improvement.

J. J. McGonagle and C. M. Vella, *Proactive Intelligence*,
DOI: 10.1007/978-1-4471-2742-0_1, © Springer-Verlag London 2012

Benchmarks: The processes and results that represent the best practices and performance for similar activities, inside or outside an organization's industry.

BI: *Business intelligence.*

Blog: Derived from combining web and log. It is a website, or part of a website, which is updated with new content from time to time. Many are interactive, meaning readers can add their own comments.

Boston Consulting Group matrix: A chart that had been created for the Boston Consulting Group to help companies analyze their business units or product lines. This helps the company allocate resources and is used as an analytical tool in brand marketing, product management, strategic management, and portfolio analysis. Best known for its terms for each quadrant of the matrix: Cash Cows, Question Marks, and Stars.

Business Intelligence: An old term for CI. Also used in knowledge management (KM) to describe the product of KM activities.

Cached: On the Internet, storing websites. Used for research on modified or deleted websites.

CI Cycle: The process of establishing CI needs, collecting raw data, processing it into finished CI, and distributing it to the end users (who then uses it). It also includes feedback among the various phases.

CI: Competitive intelligence.

Closing the Loop: Having individuals whom you interview or try to interview referring you back either to secondary resources or to individuals whom you have previously tried to interview or interviewed. Also known as Tail Chasing.

CMP: Crisis Management Planning or Plan.

Competitive Benchmarking: Involves using CI techniques to develop data on competitors which is then used for benchmarking. Differs from other forms of benchmarking in that the target, your competitor, is not cooperating in the project, and, in fact, is unaware of the project at all. Also known as Shadow Benchmarking.

Competitive Intelligence: CI consists of two overall activities. First, the legal and ethical use of public sources to develop data (raw facts) on competition, competitors, and the market environment. Second, transformation, by analysis, of that data into information (usable results).

Competitive Scenario: An analysis of what one or more competitors can be expected to do in response to changes in market and other conditions affecting their activities. Based on a profile of each competitor, including estimations of its intentions and capabilities, and stemming from a study of its past actions and of the perceptions, style, and behavior of its present and future management. Each competitor's actions are studied against the same set of expected market conditions and changes.

Competitive Technical Intelligence: Intelligence activities that allow a firm to respond to threats or identify and exploit opportunities resulting, from technical and scientific change.

Competitor Analysis: Competitor analysis involves an assessment of the strengths and of the weaknesses of current and potential competitors. This aims at bringing all of the relevant sources of competitive analysis into one framework to support effective strategy creation, execution, monitoring, and adjustment.

Competitor Array: Matrix produced by defining your industry, defining who are your competitors, determining who your customers are and what benefits they expect, and identifying the key success factors in your industry. The key success factors are ranked by giving each a weighting. Sum of all weightings must add up to 1.0. Then rank each competitor on each key success factor and multiply each cell in the matric by the factor weighting.

Competitor Intelligence: Another term for CI.

Consumer Insights: A separate corporate function tasked with proving discoveries that can lead to specific market opportunities, based on a deep and current knowledge of a firm's consumers and markets.

Corporate Security: A process is aimed primarily at protecting and preserving all corporate assets, both tangible and intangible. Typically it operates to set up protections for assets (such as databases or automobiles), to determine potential threats, and to provide a barrier. In some corporate security operations, a special focus is on the identification and protection of Trade Secrets.

Crisis Management Program: A formal method of developing, updating and implementing crisis management plans.

CTI: Competitive technical intelligence.

Customer: A person or organization, internal or external, which receives or uses outputs from one group or division. These outputs may be products, services, or information. In CI, often known as the decision-maker or end-user.

Data Mining: The process of sifting through massive amounts of data (in computer readable form) to reveal intelligence, hidden trends, and relationships between customers and products.

Data Warehousing: The ability to store large amounts of data by specific categories, so that it can easily be retrieved, interpreted, and sorted (Data Mining) to provide useful information, typically about customers and products.

Data: Raw, unevaluated material. Data may be numeric or textual. It is the ultimate source of information, but becomes usable information only after it has been processed and analyzed.

Decision-maker: In CI, the individual or group for which an intelligence evaluation is prepared. Also known as the end-user.

Disaggregation: In handling data, breaking up aggregated data into its component parts.

Disinformation: Incomplete or inaccurate information designed to mislead others about your intentions or abilities. When used in the arena of international politics, espionage, or intelligence, the term also means the deliberate production and dissemination of falsehoods, fabrications, and forgeries aimed at misleading an opponent or those supporting an opponent.

Displacement: In CI, where expected data is replaced by unforeseen data.

Distribution channel: A path through which goods and/or services flow in one direction, that is from the vendor to the consumer, and payments generated by them flow in the opposite direction, that is, from the consumer to the vendor.

Economic Espionage Act of 1996: A U.S. (federal) criminal statute, P.L. 104-294 of October 11, 1996, which criminalizes, at the federal level, the misappropriation of trade secrets. It provides additional penalties if such misappropriation is conducted by foreign entities.

EEA: Economic Espionage Act of 1996.

Elicitation: In CI, apparently ordinary conversation which is skillfully aimed at drawing out key data.

End users: Persons or organizations that request and use information obtained from an online search or other source of CI. Also known as Decision-maker.

Environmental Scanning: Study and interpretation of political, economic, social, and technological events/trends that influence a business, an industry or the market.

Espionage: Either the collection of information by illegal means or the illegal collection of information. If the information has been collected from a government, this is a serious crime, such as treason. If it is collected from a business, it may be a theft offense.

False Confirmation: When a piece of data appears to come from multiple independent sources, but in fact is coming from only one source.

FOIA: Freedom of Information Act.

Fraud: An act that involves distributing erroneous or false information with intent to mislead or to take advantage of someone relying on that information.

Freedom of Information Act: US (federal) statute, 5 USC sec. 552. Provides for access to federal government records by the public. Also used generically for all open records, public information statutes.

Gaming: An exercise that has people either acting as themselves or playing roles in an environment that can be real or simulated. Games can be repeated but cannot be replicated, as is the case with simulations and models. Also known as War Gaming or Scenario Playing.

Historic Data: Data that covers a long period of time. It is designed to show long-term trends, such as gross sales in an industry over a 5-year period. This may include projections made covering a long period of years.

Home Page: An Internet site owned by an entity that permits the user to obtain and view information provided by the owner.

HR: Human Resources.

Information: The material resulting from analyzing and evaluating raw data, reflecting both data and judgments. Information is an input to a finished CI evaluation.

Intelligence: Knowledge achieved by a logical analysis and integration of available information data on competitors or the competitive environment.

IP: Intellectual Property.

ISP: Internet Service Provider.

IT: Information technology.

KM: Knowledge Management.

KMS: Knowledge Management Systems.

Knowledge Management: The combination of Data Warehousing and Data Mining, aimed at exploiting all data in a company's possession.

Life Cycle: The time it takes a product/service to move from being new and revolutionary, through becoming established and evolutionary, to becoming passé and a commodity.

Linkage: In CI, the connection between two or more difference data sources, enabling a research to move from one to another.

M&A: Mergers & Acquisitions.

Macro-level Data: Data of a high level of aggregation, such as the size of a particular market or the overall rate of growth of the nation's economy.

Market Intelligence: Intelligence developed on the most current activities in the marketplace.

Marketspace: A term usually used in the IT business where information and/or physical goods are exchanged, and transactions take place through computers and networks. Also sometimes used as a synonym for Niche.

Micro-level data: Data of a low level of aggregation, or even unaggregated data. This might be data, for example, on a particular competitor company's profit margin for one product.

Mirror-imaging: Assuming that your adversary can or will do what you could or would do in its place.

NGO: Non-Government Organization.

Niche: When used in marketing, it is a portion of the market on which a specific product and/or service is focusing.

Nielsen: The Nielsen Company, which provides consumer packaged goods analytics and consulting services.

Open Source Intelligence: A form of intelligence collection management that involves finding, selecting, and acquiring information from publicly available sources and analyzing it to produce actionable intelligence. It generally does not involve significant primary intelligence research.

Patent Mapping: The visualization of the results of statistical analyses and text mining processes applied to patent filings. You can use bibliographic data to identify which technical fields particular applicants are active in, and how their filing patterns and IP portfolios change over time.

Primary research: Research seeking original data. Usually refers to interviewing people or developing original data.

Public domain: With respect to IP, a work is in the public domain if the IP rights have expired, if the IP rights have been forfeited, or if they are not covered by IP rights in the first place.

Pushback: A CI analyst talking to you about CI needs does not merely listen to what you say, write it down, and arrange to deliver it in 3 weeks. The analyst asks, and re-asks, questions about what you want, why you want it, in what form you want it, when you want it, and what you will do with it.

Qualitative: In measurements, this refers to evaluating the basic nature or attributes of a process or event.

Quantitative: In measurements, this refers to the accurate measurement of the components or consequences of a process or event.

R&D: Research & Development

Reverse Engineering: Discovering the technological principles of a device, object, or system through analysis of its structure, function, and operation. It often involves taking something apart.

SBU: Strategic business unit.

Scenario Playing: See Gaming.

SCIP: Strategic & Competitive Intelligence Professionals. Formerly known as the Society of CI Professionals.

SEC: US Securities and Exchange Commission.

Secondary research: Research involving the summary, collation and/or synthesis of existing research and data. Sometimes called "desk research".

Shadow Benchmarking: See Competitive Benchmarking.

Strategic intelligence: CI provided in support of strategic, as distinguished from tactical, decision-making.

Supplier: A company or person that provides inputs to tasks or jobs, whether inside or outside of the company.

Supply chain: A system of organizations, people, technology, activities, information, and resources involved in moving a product and/or service from the supplier to the customer.

Surveillance: A continuous and systematic watch over the actions of a competitor aimed at providing timely information for immediate response to the competitor's actions.

SWOT: Strength–Weakness-Opportunity-Threat analysis. SWOT analysis is a strategic planning method used to assess the Strengths, Weaknesses, Opportunities, and Threats (SWOT) involved in a project. It involves specifying the objective of the project and then identifying both the internal and external factors that support or impede obtaining that objective.

SymphonyIRI: SymphonyIRI Group, Inc., which provides information, analytics, and business intelligence solutions for consumer packaged goods, retail, environment, and healthcare companies.

Tail chasing: See Closing the Loop.

Target: A specific competitor, or one or more of its facilities, SBUs, or other units.

Technical Intelligence: Identifies and exploits opportunities resulting from technical and scientific changes as well as to identify and respond to threats from such changes.

Time to market: The length of time it takes from a product/service being conceived until it is available for sale.

Trade Secret Protection: Use of contracts, civil litigation, and criminal prosecution, under both state and federal law, to prevent trade secrets from being used by competitors.

Uniform Trade Secrets Act (UTSA): A model law, drafted by the National Conference of Commissioners on Uniform State Laws, dealing with the civil penalties for misappropriation of trade secrets. Last amended in 1985, it has been passed, in one form or another, in 45 states and 3 territories.

USPTO: US Patent and Trademark Office.

UTSA: Uniform Trade Secrets Act.

War Game: See Gaming.

WIPO: World Intellectual Property Organization, a specialized agency of the United Nations.

Chapter 2
What is Competitive Intelligence and Why Should You Care about it?

2.1 What is Competitive Intelligence?

There are a lot of definitions of Competitive Intelligence (CI), or CI as we will call it. This is the one we prefer:

> Competitive Intelligence (CI) involves the use of public sources to develop data on competition, competitors, and the market environment. It then transforms, by analysis, that data into [intelligence]. Public, in CI, means all information you can legally and ethically identify, locate, and then access (McGonagle and Vella 2002, p. 3).

CI is also called by a lot of other names: competitor intelligence, business intelligence, strategic intelligence, marketing intelligence, competitive technical intelligence, technology intelligence, and technical intelligence. The most common difference among them is that the targets of the intelligence gathering differ. However, what those who are developing it all do is essentially the same:

- They identify the information that a decision-maker needs on the competition, or the competitive environment;
- They collect raw data, using legal and ethical means, from public sources;
- They analyze that data, using any one of a wide variety of tools, converting it into intelligence, on which someone can take action ("actionable"); and
- They communicate the finished intelligence to the decision-maker(s) for their use.

To understand CI, you must first clearly understand what is meant by "public", that is, where the raw data you will need is located. The term is to be taken in its very broadest sense–it encompasses much more than studies that the US Department of Commerce releases or what you can find reported in *The Chicago Tribune*. "Public" in CI is not equivalent to published; it is a significantly broader concept.

In CI, public encompasses all information you can legally and ethically identify, locate, and then access. It ranges from documents filed by a competitor as a part of a local zoning application to the text of a press release issued by a competitor's marketing consultant describing its client's proposed marketing strategy,

J. J. McGonagle and C. M. Vella, *Proactive Intelligence*,
DOI: 10.1007/978-1-4471-2742-0_2, © Springer-Verlag London 2012

where the marketing firm also extols the virtues of its contributions to the design of a new product and the related opening of a new plant. It includes the web-cast discussions between senior management and securities analysts, as well as the call notes created by your own sales force.

Please keep in mind that CI is not just aggregating the results of a Google.com online search. That is collecting data. Admittedly, you do have to determine what you are searching for before you start, but using Google (or any other search engine—we are not playing favorites here) is just a way of picking out potentially interesting bits of data to look at from billions of available bits of data. The vast production of search engines is a classic illustration of the fact that data is not the same as intelligence. Intelligence is refined from data and is actionable. It all too often gets lost in a blizzard of raw data.

The CI process is usually formally divided by CI professionals into five basic phases, each linked to the others by a feedback loop. We are describing them to you because some, but not all, of what you will do includes some of these phases. Also, when you read further about CI, you will often find authors referring to the "CI Cycle". These phases, making up what CI professionals call the CI cycle, are—

- *Establishing the CI needs.* This means both recognizing the need for CI and defining what kind of CI the end-user needs. It entails considering what type of issue (strategic, tactical, marketing, etc.) is motivating the assignment, what questions the end-user wants to answer with the CI, who else may also be using the CI, and how, by whom and when the CI will ultimately be used.
- *Collecting the raw data.* First, a CI professional translates the end-user's needs into an action plan, either formally or informally. This usually involves identifying what questions need to be answered, and then where it is likely that he/she can collect the data needed to generate the answers these questions. The CI professional has to have a realistic understanding of all significant constraints, such as time, financial, organizational, informational, and legal. Then he/she can identify the optimal data sources, that is, those that are most likely to produce reliable and useful data, given the goal and the constraints. From there, the collection begins, both of secondary and primary data.
- *Evaluating and analyzing the raw data.* In this phase, the data that was collected is evaluated and analyzed, and is transformed into useful CI. That may be done by the person doing the collection or by a separate CI analyst. In practice, there are always two ways in which analysis is used in the entire process. The first is the use of analysis to make a selection, such as deciding which of a dozen news articles is most important to read. The second is the use of analysis to add value to one or more pieces of data. That would mean, for example, adding a statement to a summary of an article indicating *why* and *how* its contents are important to the end-user. While CI analysts provide both types of analysis, end-users most frequently only regard the latter process as really being analysis. Of course this is not true. If you do not use some analysis during the collection process, you will waste hours of time collecting useless information that takes you nowhere.

Fig. 2.1 CI Phases linked by feedback loops

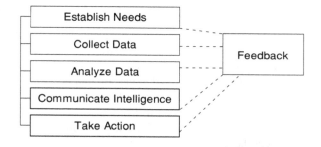

- *Communicating the finished intelligence.* This involves preparing, and then presenting, the results in a usable format and in a timely manner. The CI may have to be distributed to those who asked for it and, in some cases, to others who might also profit from having it. That secondary distribution is not as common as it could be. One study sadly notes

> [m]ore than 70 percent of employees report [this kind of competitive] knowledge is not reused across the company (Korn/Ferry 2000).

The final form of the CI, as well as its timeliness and maintaining its security, are all important aspects of its communication.

- *Taking Action.* This means using the end-user actually uses the CI in decision-making. The CI may be used as an input to decision-making, or it may be the first of several steps in an overall assessment of, for example, a new market. The decision of how and when it is used is made by the end-user, not by the analyst (Fig. 2.1).

This is all very pedantic, but you should focus on the fact that *when CI professionals, consultants, academics and the military view the CI cycle, they presume that either (a) you are doing all of this for someone else, or (b) someone else is doing this for you.*

That assumption works very well for some large businesses, with separate, free-standing CI teams available to some decision-makers, but what about all of the rest of us? *The fact is that we all need some CI to do our jobs better.* If you do not think that you need CI, you are wrong. There is virtually no commercial context in which CI cannot be a critical addition to the process of beginning a business, growing it, succeeding in it, or even of just surviving.

Ask yourself this: is my firm so successful, so entrenched, and so well-funded, that there is nothing in the competitive environment that can impact it? Of course not! What would happen if your biggest competitor suddenly and unexpectedly went out of business? Or, what if a smaller competitor was acquired by a firm 50 times your size? What if your firm's most important supplier decided to cut out the middle man—you—and became a competitor? Knowing how to develop proactive CI might at least remove the "unexpected" from these unexpected scenarios. These and hundreds more critical events are places where effective CI, which can give you an early warning, can be a powerful tool for you.

So what does the CI cycle look like for you? It looks *very, very different*:

- *Establishing your CI needs.* This means recognizing that *you* need CI. It means considering what type of issue (strategic, tactical, marketing, etc.) is motivating you, and what questions you want to answer with the CI. For you, it also means determining what the *real* issues are that you have to deal with, those underlying your starting questions. We will give you pointers so that you can focus quickly on what you really need to know.
- *Collecting the raw data.* You have to identify where it is likely that you can collect the data you need to answer your questions. You need to have a realistic understanding of any significant constraints, such as time, financial, organizational, informational, and legal. For example, you may not be able, or willing, to call someone at a competitor's consumer information center for an answer to a simple question like "Where is this new product being sold?" Only then can you identify the data sources that are available to you and that are most likely to produce reliable, useful data.
- *Evaluating and analyzing the raw data.* In this phase, the data you are collecting is evaluated and analyzed—by you—and you transform it into CI for your own use.
- *Communicating the finished intelligence.* This is not applicable (unless you are talking to yourself).
- *Taking Action.* This involves using your own CI in your decision-making.

That all seems to represent pretty simple changes to the classic CI cycle, but embedded within these changes is a critical difference: you have no one helping you with your CI work, while the CI cycle assumes that there are at least two people working together. In particular, you have to use methods, which we will tell you about, to refine your own first questions into a statement of what you really need. You will have to do the data collection yourself, which almost always means you have less time to devote to it than would a full-time CI professional. And then you have to do the analysis yourself. That can be much more difficult than you think, but we will help you there too. We will give you specific guidance on how to do this and to compensate, as much as is possible, for the loss of an experienced intermediary. In addition, by doing this, you are by-passing the CI disconnect—you can make sure that your CI will be used in your decision-making.

2.2 Why Should You Care about Competitive Intelligence?

Consider the findings of a McKinsey study that asked executives how their firms responded either to a significant price change by a competitor or to a significant innovation by a competitor:

> A majority of executives in both groups [across regions and industries] say their companies found out about the [significant] competitive move too late to respond before it hit the market (McKinsey 2008).

Let us look at it from a more positive angle. CI does provide value, even though virtually all evidence of the value and impact of CI is, to date anecdotal or deal with indirect assessments. Here are a few of the key ones that should help you feel better:

- In the early 1990s, a study of the packaged food, telecommunications and pharmaceutical industries reported that organizations that engaged in high levels of CI activity show 37% higher levels of product quality, which is, in turn associated with a 68% increase in business performance (Jaworski and Wee 1993).
- In the mid 1990s, NutraSweet's CEO valued its CI at $50 million (about $72 million in 2010 dollars). That figure was based on a combination of revenues gained and revenues which were "not lost" to competitive activity (Flynn 1994).
- A more recent PricewaterhouseCoopers' study of "fast growth" CEOs reported that "virtually all fast growth CEOs surveyed (84%) view competitor information as important to profit growth of their company" (PricewaterhouseCoopers 2002).

2.3 What Kinds of CI are There?

Today, CI, as it is practiced, is often divided into different, but overlapping, types. We feel comfortable with dividing it into strategic, competitor, tactical, and technical. The terms are simple, and communicate how the CI is intended to be used. They each share the common concept of where the data comes from and a common toolbox to help in its analysis. To you, the differences are sometimes important in terms of what questions you can seek to answer, where you can best look for data, and which analytical tools you should use.

2.3.1 Strategic Intelligence

2.3.1.1 What is Strategic Intelligence?

Strategic intelligence is CI supporting strategic, as distinguished from tactical, decision-making. This means providing higher levels intelligence on the competitive, economic and political environment in which you firm operates now and in which it will operate in the future.

2.3.1.2 Who and What Does Strategic Intelligence Help?

Strategic intelligence typically is used by senior managers and executives who make and then execute overall corporate strategy. Its most common applications are in the development of the following:

- Long-term (3–5 year) strategic plans
- Capital investment plans
- Political risk assessments
- Merger and acquisition, joint venture, and corporate alliance policies and plans
- Research and development planning.

2.3.1.3 What Does Strategic Intelligence Focus on?

Strategic intelligence usually focuses on the overall strategic environment. A firm's direct competitive environment and its direct competitors are, of course, included in that focus. It should also include its indirect competitors. In addition, strategic intelligence should develop CI on the long-run changes caused by, as well as affecting, all of the forces driving industry competition, including:

- Suppliers
- Customers
- Substitute products or service, and
- Potential competitors.

You conduct strategic CI analysis when you must focus on many critical factors, such as technology trends, regulatory developments and even political risks that, in turn, effect these forces.

Strategic intelligence's focus is less on the present than it is on the past, and is primarily on the future. The time horizon of interest typically runs from 2 years in the past to 5 or even 10 years in the future.

- In terms of an interest in the past, you will be collecting and analyzing data so that your firm can evaluate the actual success (or failure) of its own strategies and of those of your competitors. This, in turn permits you better to weigh options for the future. You are looking to the past to learn what may happen in the future.
- With respect to the future, you are seeking a view of your firm's total environment: competitive, regulatory and political. As with radar, you are looking for warnings of impending problems, and alerts to upcoming opportunities—always in time to take needed action.

2.3.2 Competitor Intelligence

2.3.2.1 What is Competitor Intelligence?

Competitor intelligence focuses on competitors, their capabilities, current activities, plans, and intentions.

2.3.2.2 Who and What Does Competitor Intelligence Help?

Competitor intelligence is most often used by strategic planning operations or by operating managers within strategic business units (SBUs). It may also be useful to product managers, as well as to those involved with product development, new business development, and mergers and acquisitions.

2.3.2.3 What Does Competitor Intelligence Focus on?

Competitor intelligence usually helps you answer a wide variety of key business questions, including ones such as these:

- Who are our competitors right now?
- Who are our potential competitors?
- How do our competitors see themselves? How do they see us?
- What are the track records of the key people at our competitors? What are their personalities? What is the environment in their own company? What difference do these people make in terms of our ability to predict how these competitors will react to our competitive strategy?
- How and where are our competitors marketing their products/services? What new directions will they probably take?
- What markets or geographic areas will (or won't) be tapped by our competitors in the future?
- How have our competitors responded to the short and long-term trends in our industry in the past? How are they likely to respond to them in the future?
- What patents or innovative technology have our competitors or potential competitors recently obtained or developed? What do those changes and innovations mean to us?
- What are our competitors' overall plans and goals for the next 1–2 years in the markets where they currently compete with us? What are their plans and goals for their other firms and how will those affect the way they run their business competing with us?

Competitor intelligence's time horizon typically runs from 6–12 months in the past to 1–2 years in the future.

2.3.3 Market Intelligence

2.3.3.1 What is Market Intelligence?

Market intelligence is focused on the very current activities in the marketplace. You can look it as it as the qualitative side of the quantitative data research you have conducted in many retail markets.

2.3.3.2 Who and What Does Market Intelligence Help?

The primary users of market intelligence are usually the marketing department, market research, and the sales force. To a lesser degree, market intelligence serves those in market planning by providing retrospective data on the success and failure of their own sales efforts.

2.3.3.3 What Does Market Intelligence Focus on?

Market intelligence's focus is on sales, pricing, payment and financing terms, promotions being offered and their effectiveness. Market intelligence's time horizon typically runs from 3–6 months back to no more than 6 months in the future. Some of the time, however, the horizon is actually measured in terms of weeks, or even days, rather than months.

2.3.4 Technical Intelligence

2.3.4.1 What is Technical Intelligence?

Technical intelligence permits you to identify and exploit opportunities resulting from technical and scientific changes as well as to identify and respond to threats from such changes.

2.3.4.2 Who and What Does Technical Intelligence Help?

Technical intelligence is particularly useful if you are involved with your firm's research and development activities. Using basic CI techniques, those practicing technical intelligence now often can determine the following:

- Competitors' current manufacturing methods and processes.
- A competitor's access to, use of, and dependence on, outside technology, as well as its need for new technology.
- Key patents and proprietary technology being used by, being developed by, or being acquired by, competitors.
- Types and levels of research and development conducted by competitors, as well as estimates of their current and future expenditures for research and development.
- The size and capabilities of competitors' research staff.

2.3.4.3 What Does Technical Intelligence Focus on?

Technical intelligence has a slight overlap with both competitor and market intelligence, particularly with respect to its interest in suppliers and customers. However, instead of dealing with market trends, Technical intelligence is usually focused on technology trends and scientific breakthroughs. Technical intelligence projects can develop information on opportunities for your firm as well as threats to the firm. Technical intelligence's time horizon typically runs from 12 months in the past to 5+ years in the future.

2.4 What is NOT Competitive Intelligence?

Unfortunately, the wide variety of names that those of us in CI have used has caused, and probably will continue to cause, confusion between CI and other knowledge-based activities. The most frequent areas of confusion are with environmental scanning, business intelligence, knowledge management, and market/ quantitative research.

Why do you care? Because when you do your own CI, you want to make sure that someone does not tell you that "We are already doing that sort of thing here, so there is no reason for you to have to do it" when they are actually referring to something else. Also, when you read more about CI, you want to make sure you are reading the right books.

2.4.1 Environmental Scanning

As the term "environmental scanning" is used today, its emphasis is on the future, not the present or the past. In addition, its stress is generally heavily on data acquisition to generate an early warning of problems, rather than on subsequent analysis, to support a wide range of decision-making (See Abreu and Castro 2010). However, to add to the confusion, some CI professionals may use the term environmental scanning to give a broader mission to their research efforts than would be provided by just calling it plain old CI.

2.4.2 Business Intelligence

"Business intelligence" is a particularly difficult term to deal with. At one time, this term was actually used by some CI professionals to describe CI in a very broad way, and to describe only intelligence provided in support of corporate strategy by others. Now its use seems to have been fully co-opted by those involved with data management and data warehousing. There, it can refer to:

- The software used to manage vast amounts of data,
- The process of managing that data, also called data mining, or
- The output of either of the first two.

In any case, virtually all of the reported applications and successes of business intelligence deal with processes which are internally-oriented, from process control to logistics, and from sales forecasting to quality control. The most that can be said of it its relationship to intelligence is that

> data mining and related techniques are useful tools for some early [terrorism intelligence] analysis and sorting tasks that would be impossible for human [intelligence] analysts. They can find links, patterns, and anomalies in masses of data that humans could never detect without this assistance. These can form the basis for further human inquiry and analysis (DeRosa 2004, p. 6).

2.4.3 Knowledge Management

First, most knowledge/data management systems (KMSs) are essentially *quantitative* in focus, while CI, as a discipline is most often *qualitative* in focus.

Second, those conducting CI often need to be able to access the people who provided the data as well as the data. Why? Data gives only the past; people can help you to see into the future. But, again, most KMSs are keyed to storing and manipulating data. They rarely allow precise identification of a human source(s), much less information on obtaining immediate and direct access to him/her.

Third, most KMSs are not set up to capture data on anything that does not involve the firm itself. Yet firm personnel, from the CEO down, interface daily with customers, from whom information on competitors can be developed, as well as with suppliers, distributors and the like. All of those in the supply chain, for example, can be powerful sources of useful CI data. But KMSs typically do not provide access to them.

Fourth, the sales force, potentially a very powerful source of data in support of CI, is rarely involved with KMSs and related efforts. The sales force often sees any initiative that does not produce immediate sales opportunities as a distraction or even irrelevant. Yet, those firms who can tap into the sales force as a part of the CI process have found significant benefits for both sides of the transaction.

Fifth, few, if any, KMSs provide access to current information on current employees. And knowing which employees are members of what professional association, which have gone to what trade shows, where they worked before and what they did there is something of great value to many seeking to gather CI data.

Sixth, KMSs do not record decision-making and the history of decisions. KMSs have the potential, as yet untapped, to serve as the repositories of enterprise history, at the strategic as well as product and service levels. But, we know of none which, for example, access previous strategic plans, much less retrospective evaluations of their success, and, more importantly, their failures.

2.4.4 Market Research and Quantitative Research

While CI does use some quantitative methods in conducting its analysis, it does not do so to the degree that most quantitatively-oriented researchers do. To draw a somewhat imprecise line, market research focuses on competitors and the firm's own interface with its customers on an historic and real-time basis. CI focuses on a broader horizon, including potential competitors, the supply and distribution chains, and research and development. In addition, its perspective is most often forward-looking. To play off an advertising slogan, CI seeks answers to questions like "Where do they want to go tomorrow?" Finally, CI, because it is forward-looking, is heavily qualitative (stronger, weaker) in comparison with more market research and qualitative research (15.3%, 2.3 million units).

References

Abreu PGF, de Castro JM (2010) Are we blinded by the "traditional" intelligence cycle? Competitive Intell 13(3):18–26

DeRosa M (2004) Data mining and data analysis for counterterrorism. Center for Strategic and International Studies, Washington March

Flynn R (1994) NutraSweet faces competition: the critical role of competitive intelligence. Competitive Intell Rev 5(4):4–7

Jaworski B, Wee LC (1993) Competitive intelligence: creating value for the organization—final report on SCIP sponsored research. The Society of Competitive Intelligence Professionals, Vienna

Korn/Ferry International and the University of Southern California's Center for Effective Organizations at the Marshall School of Business (2000) Strategies for the knowledge economy: from rhetoric to reality. http://www-marshall.usc.edu/assets/046/9699.pdf. Accessed 22 June 2011

McGonagle JJ, Vella CM (2002) Bottom line competitive intelligence. Quorum Books, Westport

McKinsey & Co (2008) how companies respond to competitors: a McKinsey global survey. http://www.mckinseyquarterly.com/How_companies_respond_to_competitors_2146. Accessed 11 Oct 2011

PricewaterhouseCoopers (2002) One-third of fast-growth CEOs place higher importance on competitor information than a year ago. http://www.barometersurveys.com/vwAllNewsByDocID/03295DF410AE990A85256BA6000013AC/index.html. Accessed 11 Oct 2011

Chapter 3
How Can You Use CI in What You Do?

3.1 What Can You Do with CI Once You Have it?

The key for you is to develop only Competitive Intelligence (CI) that *you* can *use*. If you do not use it, there is no point in doing all the work you did to collect the raw data and to analyze it. Above all, remember why you developed this CI in the first place: to make a decision and then to take action.

3.2 Best Practice Examples

Where can you use CI? A better question is where cannot you use it? Consider the fact that the Baldrige Performance Excellence Program tells potential applicants to start the evaluation process by completing an Organizational Profile. This Profile requires a firm to describe its Competitive Environment, including its competitive position and where it gets its competitive data. (National Institute of Standards and Technology 2011, pp. 1, 6) The Baldrige process puts CI in a primary position.

What follows are some specific best practice examples. Keep one thing in mind—you can use your CI skills anywhere. CI is primarily used to determine who your competitors are, where they are currently positioned, or most importantly, where they are going tomorrow.

Once you understand that, we will move into defining your needs, locating the data you need, and analyzing it to make it actionable for you.

3.2.1 In Marketing and Product Development

In a large company that has a marketing team and a product development arm, either separate or tightly linked, CI is usually done in these units or by a separate CI unit and reports to the marketing unit itself. When you do your own CI in these units, you have much more control and your company can benefit in a variety of ways.

For example, in 2004, the Society of Competitive Intelligence Professionals (SCIP) identified the most effective ways that marketing units used CI (Society of Competitive Intelligence Professionals 2004). The top-rated were as follows:

- Decision support
- Market monitoring
- Identifying market opportunities
- Market plan development
- Market plan input
- Investigating market rumors
- New product development
- Anticipating competitor initiatives
- Identify alliance partners
- Anticipating technology changes
- Investment prioritization
- Identifying competitor intangibles.

Consider these specific cases where you can use CI:

- When it comes to new product development, you can help to identify and track existing and potential competitive products. You can use a combination of access to the US Patent and Trademark Office (to check for new trademark filings), followed by a search of patent application publications, and then for new domain name filings, and then go to job boards, advertising, and press releases to quickly check on potential new product rollouts.
- When preparing for a launch of a new product, CI can be used in other ways:

  When [Lexis–Nexis] launched our Web products for the business markets…we used CI to decide how to position it and to learn what were the key elements in the product that customers would like to have….[In another case, we were] not looking at what our traditional competitors do, but looking at what *our nontraditional competitors are doing*…what they're working at in providing information tools to companies. (Gieskes 2001, p. 79)

- In marketing, you can track the production of new products or services, such as those to be produced in a new plant still under construction (Of course, your CI activities should have already alerted you to its pending construction). Knowing how the construction is actually going enables you to know when you will have to respond in the market.
- In new product development, you can analyze the product development process at competitors, and benchmark your own program against world class product development programs, without going through a full cooperative benchmarking exercise. What you learn from each will help you to improve your own program. Take for example this case:

  Colgate read the signals appropriately regarding a competitor's product launch of a new toothpaste container. The company pre-empted the threat by launching a price cut on existing products (get 3 products free) before the competitive launch. It thus bought time enough to launch its own competing product. (Global Intelligence Alliance 2006, p. 11)

- In marketing and product development, you can track changes in marketing and executive personnel at competitors, knowing that new personnel can mean forthcoming changes in direction. By knowing their backgrounds, you can more accurately determine why they are doing, what they are doing, and how successful they might be.
- What kind of value can your use of CI provide here? Let us give you an example from our practice:

> A food company sales person told a product manager that she had seen a new product by a small, private competitor, in the grocery store. Now that bit of data could have stopped there. But it did not. The product manager started a process, digging into this anomaly, that eventually uncovered a very under-the-radar effort by the competitor to build a new plant. This was to be followed by a roll out of this new product on a national basis. Solely because the product manager asked "What does this mean for us?" the company had advance warning of a major competitive change coming up.

3.2.2 In Strategy and Strategic Planning

All too often, strategy development proceeds with a rigid focus on the firm doing the planning, but with less, or even no, CI on its competition. The worst case is that those developing a strategy assume (always a dangerous word in CI!) that their competitors will continue to do next year what they have done this year and last year (assuming they even know that). In other words, they completely ignore the fact that these competitors will have to respond to whatever strategy that their own firm is developing. And that they will then have to respond to the competitors, etc.

CI's importance in the planning process is recognized by the Association for Strategic Planning (ASP). ASP's annual Richard Goodman Award application specifically asks contestants "How do you use competitive data?" (Association for Strategic Planning, 2011), not "Do you use competitive data?"

Using CI in strategy development and strategic planning prepares you and your firm:

> [One of the CI unit's responsibilities] is to help us understand if there are any competitor moves in the last quarter that would change our view of the competitive landscape and help us understand [organizations] that have the ability to come in and play a competitor role in our marketspace. (Hovis 2001, p. 91)

In addition, CI can help in predicting when competitors will undertake new initiatives, as a recent McKinsey study indicates:

> The timing of many competitors' moves can be predicted, because these moves either result from an annual planning process or are prompted by external events visible to all companies. Companies should be sure they understand the external landscape of their industry at least as well as their competitors do (and investigate when their competitors conduct their annual planning process). (McKinsey 2008)

A dangerous case is when the strategic plan is developed or evaluated by those with personal or institutional "blind spots". Those with blind spots can be easily identified by phrases they use such as "We know", "We have been doing this for [fill in the blank] years", and "We understand the market better than the competition does". In every case, the presence of these blind spots means that decisions are being made on the basis of incomplete, inaccurate, or dated information, if in fact they are made based on any real information. We will deal further with blind spots and how you can handle them later.

Another problem lies in tracking the success, or the failure, of the firm's strategic plan once it is implemented. Very few firms take the time to track how the competition is, or is not, responding to the firm's plan. If you can use CI to compare your firm's actual performance against the predicted performance by your competitors it means that next year's plan will be even better, since it will be based on both current intelligence and continuous monitoring. Here, business war gaming, discussed later on, can be a powerful way of delivering CI to many managers to test your planning before execution.

3.2.3 In Sales

A study several years ago detailed the most effective ways that sales teams said that they utilized CI:

- Determining competitor capabilities
- Understanding the full range of competitor products/services
- Finding out competitor pricing
- Understanding customer demand
- Advance notice of features/function changes
- Identifying specific customer requirements
- Relaying competitor sales approaches
- Support for making specific bids. (SCIP 2004)

Note that, in every case, the sales teams used CI that led them to carry out their mission—to make more sales.

Consider this case:

At Pergo Inc., a maker of laminate flooring, intelligence helped steal a major contract right out from under a rival's nose. When Pergo told a national retailer what it had learned from a mutual supplier–that the rival would not be able to launch a new product when it said it would–the retailer signed with Pergo instead. Says [a Pergo] Senior Product Manager..."It's not rocket science. It's basic blocking and tackling." (Lavelle 2001)

3.2.4 In Human Resources

Very few human resources departments are even aware of CI, much less using it regularly. In our practice, we have found that CI can be very useful in developing competitive benchmarking profiles on a wide variety of subjects, including the following:

- Compensation structures
- Recruiting practices, both entry level and lateral entry
- Retention practices and rates
- Employee training
- Employee promotion and review practices
- Mentoring programs
- Minority development efforts.

Also, consider this application:

> When it comes to profiling executives – and people in general - you've really got to pause up front and make a clear choice between the deep ocean or the big blue sky. With the Internet and a few hundred dollars in hand, there is almost no limit to what you might uncover about any individual. Start by listing what you need to know, then add in what you'd like to know, and get rid of what you do not want to know..... (Carpe 2003)

In addition, you and the HR department can assist others in their own CI work by helping identify and then access former employees of competitors who are now working for your firm.

3.2.5 In Scenario Playing

Scenario playing, also known as war-gaming, involves working with your executives and managers to develop responses to potential new trends, to plan marketing campaigns, etc. For these efforts to work at all effectively, your "players" must compete against other players who can offer responses from your competitors that are fact-based and realistic. And some of the best persons for that task are those employees who track your competitors or interface with your competitors on a daily basis. Using CI you develop for the exercise, they can quickly supplement their focused knowledge with broader perspectives. Without accurate information about the competition, and a realistic assessment of its current intentions, capabilities and probable responses, the best scenario playing can, and in fact will, quickly turn into a futile exercise. With it, it can be invaluable.

One variant is to take members of your executive team and have them play the competition. For this to succeed, and it can, this team must be *fully* briefed on the competition's strengths, weaknesses, history, and executives' view of the business environment. And what does this require? Good Proactive CI from you.

3.2.6 *In Mergers and Acquisitions/Divestitures*

CI is not nearly as useful during the execution of a merger and acquisition (M&A) transaction as before that stage occurs. The most common way that CI techniques can assist is in researching an M&A target before your firm makes *any* contact with it. Why?

Because once the M&A dance begins, the parties would have entered into a variety of agreements covering their research into each other (due diligence) as well as what will happen if the deal does not go through—often with a requirement that the potential acquiring firm stays out of a particular line of business, market, etc. for a period of time. This is designed to protect the acquisition target from competition once it opens itself up.

Consider this situation:

> When a leading cell phone software developer decided to explore the option of acquiring a smaller, privately held rival company, they turned to [a CI firm]. The [developer] had heard rumors that the target company was experiencing serious financial difficulties, if this were true the [firm] felt the target might be willing to sell.
>
> [The CI firm] learned that the rival company was indeed struggling financially and had in fact lost a major retail client....[F]urther primary intelligence research revealed that the target company was pursuing a new technology and had recently signed a contract with a major ODM [original design manufacturer]. The new business model and contracts represented a 200% jump in revenue.
>
> Not only did the intelligence help prevent the client from prematurely revealing their intentions, the early warning...enabled [the developer] to move quickly to counter the strategy and maintain its market share dominance. (Sedulo Group 2011)

The same is true when dealing with divestitures, as this case involving the global engineering company ABB Group shows:

> Management was evaluating the potential divestment of a certain part of a business. The perception was that the business did not seem to have a sound outlook for the future since the market growth was limited.
>
> We did our analysis and came to the conclusion that the future capital expenditures of the particular vertical industry where this business was active in, were in fact likely to increase quite dramatically. Based on this intelligence, the divestment plans were stopped. It seems now that our analysis was indeed right, the industry sector picked up and ABB is making good business in that field. (Global Intelligence Alliance 2008, p. 22)

Using CI techniques beforehand overcomes or at least avoids these limitations and problems. During that period of work, you, the researcher, can use CI techniques to dig into the target firm, its market, etc., but do this without the cooperation of the target. While this research is less perfect than that after agreements have been signed, it enables your firm to back away from a transaction early, before negotiations even begin, without being prevented from entering that market in the immediate future.

Also, once M&A agreements are signed, effective M&A teams should be prepared to use CI to look far ahead, well beyond the closing that is scheduled, to prepare the new owners for a new competitive environment, focusing particularly on how others in the market space will respond to the forth coming changes the acquiring firm will initiate.

3.2.7 In Research and Development

While those active in Research and Development (R&D) obviously conduct their own research, having you add your CI skills here can be powerful. First, you can use CI to focus not just on the research being done by competitors, but on the scope and dimensions of their R&D programs, including estimates of their current and future expenditures on R&D. Finding out that a competitor will soon be cutting back (or doubling) spending in certain research areas can help make your own program much more effective and focused. Similarly, you can better identify and then tap into research elsewhere, and avoid duplicating research already underway at institutions you can access.

Do not assume that you should be looking only for hard data for your R&D program. In many cases, you may find that all that is available is "soft," that is, non-quantifiable or largely opinion, data. In fact, in some R&D contexts, your best CI may be "soft." For example, in discussing keeping up with Japanese R&D activities, scientists at the Battelle Pacific Northwest Laboratories observed:

> [S]cientific and technological advances in Japan are often first communicated by word of mouth within the Japanese scientific community…. It is also important to recognize that many significant technical developments in Japan do not appear directly as experimental results or as business news…. Thus technical journals and databases are helpful but only supplementary. (Anon 1987, p. 7)

3.2.8 In Crisis Management Planning

While the primary focus of this planning is a crisis management plan (CMP) that is internal, the best designed CMPs also look at the past experience of competitors (and other firms) to try and anticipate probable future crises that could face your own firm. An effective CMP should also take into account what kind of a crisis could impact your firm if one of your suppliers, customers, competitors, or lenders have a major crisis of its own.

CI can (and probably should) be involved in CMPs in a variety of ways:

- Helping your firm learn from its competitors' past actions about the kinds of crises to expect and how to handle them effectively.
- Tracking competitors and your key suppliers, customers, and lenders, all of whose actions or own crises may trigger a follow-on crisis for your firm.
- Helping your firm anticipate dealing with particular crises, such as receiving trade secrets from a disgruntled employee of a competitor, or a threat to send/ sell your trade secrets to a competitor.

- Monitoring the development of a crisis in your firm, and analyzing the often incomplete, rapidly changing, and often inaccurate information at the beginning of any crisis.
- Using CI research techniques to track (and analyze) your own firm's management of a crisis during and after the fact.

3.2.9 In Benchmarking

Your CI skills can link with benchmarking efforts in two ways:

- First, your CI skills should be able to assist a benchmarking team in determining which targets are truly "class", whether that is market class, industry class, or world class. Just because a firm is big it does not mean it is good. Your extra digging can produce greater benefits.
- Second, experience in developing your own CI means you bring to a benchmarking project a skill set which is particularly useful in cases of competitive benchmarking, also known as shadow benchmarking. Once these types of benchmarking target(s) have been selected, you can use CI techniques to develop the detailed intelligence needed for benchmarking a competitor. What occurs otherwise, in the absence of such intelligence, will just be a generalized study based on internal information only. CI provides a window on the world, allowing new information and a new perspective.

3.2.10 In Reverse Engineering

Reverse engineering is intimately tied to CI, both as a consumer of your CI analysis and as a supplier of data and analysis to you when you are developing your own CI. As a consumer of CI, reverse engineering calls upon your CI skills for supportive information, such as that on marketing efforts and distribution channels, not obvious from the product (or service) being reverse engineered.

As importantly, reverse engineering efforts can often support your own CI work. For example, some firms have found that, by regularly buying a competitor's products from the same sources, they can sometimes tell at what capacity competitor factories are operating. One clue may come from the fact that each product from a particular plant often has a unique and consecutive serial number. If the output from a key plant is found to be routinely distributed through one distribution center, the serial numbers may show relative production. This means that buying the same item there each week is a way of sampling, tracking, and thus measuring the factory's output.

Another example is using reverse engineering techniques to compare your supply chain with that of your competitors. CI can enable you to compare your costs/vulnerabilities with those of your direct competitors. In one case, we were able to tell a client that a major competitor had changed its supply chain management and, for efficiency and cost reasons, had settled on a sole supplier for a key ingredient. When a problem suddenly developed with that supplier, the client was able to exploit the forth-coming shortages that the competitor would face, almost before the competitor realized the problem was upon it. You can also spot such mundane changes as making the product size smaller.

3.2.11 In Intellectual Property Protection Programs

Most corporate intellectual property (IP) protection programs involve the careful use of the varied legal protections provided by the patent, trademark, copyright, and associated laws. While the legal protection provided by these laws can be formidable, all these legal regimes have one requirement in common—the materials, ideas, concepts, invention, designs, etc. being protected must all necessarily be disclosed to the public, including to competitors, to become protected by law.

But there are more pieces to the intellectual property security issue that operate to prevent against disclosure. The most important of these is the protection that trade secret laws provide.

Trade secrets are governed by two different laws: the Uniform Trade Secrets Act (UTSA) and the Economic Espionage Act of 1996 (EEA). To trigger their legal protections, your firm must be very careful in how it handles and protects its "trade secrets". Under the equivalent state laws, a business can bring a civil suit for damages which were caused by "misappropriation" of a "trade secret". Under UTSA, and similarly under the common law of states that have not yet adopted the UTSA, each of these concepts is a carefully defined term.

While both of these legal regimes appear to be quite sweeping, there are several important concepts embodied in both the UTSA and the EEA which actually mean that their impact on protection of critical business information is limited.

- First, the information you are trying to protect as a trade secret must be able to be specifically identified. It is not enough to say that "everything here is a trade secret". To protect information, someone must first formally identify it as such.
- Second, the information involved really must *be* a trade secret. One key to determining if this is the situation is to answer the following question: has what you are seeking to protect been the subject of reasonable efforts to keep it secret? If the answer is no, it cannot be a trade secret. For example, if it is a document and it is stamped "Confidential" at the bottom of each page, you are well on the way to protecting it. On the other hand, if the same document is included in promotional presentations being given to thousands of customers, it is not a trade secret.

- And third, is the information you think you are protecting as a trade secret "readily ascertainable by proper means"? This means that the *deduction or reconstruction by proper means of what may in fact be a trade secret is not a violation of either law.* Why? Because deduction or reconstruction is not the same as misappropriation.

All of this means that *trade secret protection can be lost through disclosures, whether accidental or purposeful,* made in any of the following common contexts:

Information is revealed in the published literature, such as trade journal articles or interviews.

- Scholarly articles or technical papers are published and circulated by in-house scientists, containing information sought to be protected.
- The existence of a "secret" product is readable in the background of photographs in an annual report or PowerPoint presentation to stock analysts.
- Critical performance data is partially revealed through advertising claims.
- Disclosures of firm secrets are made by the firm through training its customers or in technical bulletins distributed to them.
- Reviews in newspapers and trade papers contain significant, previously undisclosed, product details.
- "Secret" products are displayed at a trade show or industry conference where they can be examined by attendees, including by competitors. That is more common than you think. The head of an exhibit-strategy firm has reported that

85 percent of exhibitors don't train their booth staff to identify all visitors before positioning and presenting the company's marketing and sales messages. (Kaltenheuser 2002, p. 54)

To understand how much the Internet is changing this area, consider the following description of a 2011 NY lawsuit:

The case involved a common suit to stop a former employee from using a contacts list for a new employer....In the court's view, the fact that information that was once confidential had become freely available online...had so weakened traditional state law trade secret protection, that it was no longer bound by old...precedent protecting such assets.

The new employer was instead able to show the court [in a courtroom demonstration] that it could recreate the list online, in just a few minutes, using Google, LinkedIn and Facebook. (Jaskiewicz 2011, p. 11)

So how can CI help you in running these IP programs?

- One key way is to see if something should be protected, or, to put it another way, whether or not it has already been compromised. For example, your firm may have an internal pricing guide, enabling it to develop quotations for potential customers. Using the CI techniques which you have mastered, you might decide to search the Internet to see if this document has accidently been posted by a subcontractor. If it has been posted, then you know that your firm must do something different to protect this as the trade secret laws will no longer help you because this has already been disclosed.
- Another is to keep better track of competitive IP issues, doing more than merely accessing the USPTO, WIPO, and foreign national offices. For example,

More and more in-house counsel and general practitioners are seeing competitive intelligence on intellectual property (IP) added to their responsibilities. 'It's critical that corporate counsel with new IP-related responsibilities understand how to stay on top of the latest IP filings of competitors," said [Mark Medice, national brand manager, West]'In-house counsel also should understand the country rules and related IP lawsuits, how to make the link between IP filings and related dockets, and how to perform an IP portfolio analysis." (WestBlog 2008)

3.2.12 In Corporate Security

Your knowledge of CI can be useful, even if you are working in corporate security. To over simplify, corporate security's mission is to protect your firm against external threats, usually illegal, and against internal actions that can imperil your business, like not protecting passwords.

However, some of what companies think of as the result of security breaches may only be the result of effective CI by your competitors. Think about this situation:

- A franchise firm was concerned that its competitors were stealing its new product and pricing changes before they were announced. Looking at it from a CI point of view, we found that the firm was sending these announcements to its franchisees. However, for reasons that are not important here, a few of the franchise holders also owned franchises offered by competitors. These people saw no reason not to share the new information with the managers of their (competing) franchises. And, of course, that information quickly made its way up the chain to the competitors. Once the competitors realized that the firm was making this mistake, it asked the franchise owners to send the information directly. And they did.

3.2.13 In Small Businesses

When you first decide to open your own business, you can easily find information on how to keep your books, design a business plan, file your business taxes, and even decorate your store or office. You can find information easily on almost everything, except determining who your competitors are, what they are doing, and how they can make sure your new business has a brief lifespan.

Unless you buy into a well-known franchise business that controls where you are able to open your new business, there is absolutely nothing to stop you from opening your new retail shop directly across the street from where another competitor will be soon opening its next store. In a boutique business, this may not be the best of ideas. If you had done good CI at the beginning, you would have avoided this.

CI is a necessary component of opening any business and should never be ignored. This is as true for a small business as for a major corporation. You may not be able to hire a competitive intelligence firm to do your CI work, so you will have to do it yourself. But, with some time and effort, you can do it and you must do it. This will give your business a much better chance of success. And you should start your own CI well *before* you open your business.

The beauty and benefit of CI is that it is not a one-time opinion, such as you get from a consultant or other professional, but it is factual information about your competitors you have developed, and thus can replicate in the future. You can adjust your business plan because you are working with actual information, as opposed to opinion or gossip. As an example, if you do your CI properly and diligently, you can determine what form of advertising is working best for your competitor and, possibly, what it are paying for it. This can help you immeasurably in designing and executing your own new advertising campaign.

Take this example of a business getting ready to open:

> [An individual who owned a restaurant] had heard word that another restaurant planned to open down the block. He needed to find out more…. [He made] trips to the zoning board, the fire marshal's office and the department that regulated building codes. "All for free," he says, "we found plans for this other restaurant." The competitor's public application for the fire department permits revealed the occupancy level they were seeking. Building code plans gave information on plumbing, electrical work, number of ovens and the layout of the restaurant.
>
> By analyzing that information, the business owner determined what type of restaurant was coming and what kinds of food it would serve. He also made a pre-emptive strike with his own business. By checking with the parking bureau, he found that the competitor had applied for a valet permit. (Roop 1998)

The same is true once your small business is up and running:

> The owners [of a dude ranch] keep track of competitors through meetings of the Arizona Dude Ranch association, and by watching for ads in key print publications. Seasonal employees, who frequently have experience working on other dude ranches, are encouraged to share their experiences. (Miller 2000, p. 230)

In Professional Offices

There is a common misconception that professionals, such as lawyers, accountants, and physicians, should not worry about the competition. The rationale is that they should focus on their clients, on serving them as well as they can. This ignores an underlying issue—regardless of how professional you are, you still have to get and then retain clients. And you have to succeed in your business to keep serving them as a professional.

Let us take an example of the situation that has faced small accounting firms in the past decades:

- On the one hand, the tax laws have become increasingly complex each year. According to one source, the Internal Revenue Code alone has grown from about 400 pages in 1913 to over 67,000 pages today. (Burgess 2011). This would seem to generate a fantastic opportunity for accountants to expand their tax preparation practices continuously.

- On the other hand, taxpayers can now use individual tax preparation firms or software for their increasingly complex returns. This would seem to cause problems for accountants' tax preparation practices.

What has actually happened? One study indicates that not only has the total revenue in the tax preparation market fallen, but the revenue per firm has done so as well. (AnythingResearch 2011).

Were the sole practitioners and smaller firms aware of this? Did they track and analyze the future plans of the tax preparation software firms to see that they would eventually have to move from individual returns to small business returns? Did they follow the national tax preparation firms and notice that their expansion turned to contraction due to rules that restricted the profitability and even availability of their "refund anticipation" loans? If they used CI, they would have.

The same issues can face physicians. For example, in some communities, local hospitals are gradually adding local physicians to group practices operating under the umbrella of the hospitals. When did local physicians first see that starting? More pointedly, when could they have seen it coming? What will it mean to them? If one hospital does it, will the others follow? If a hospital is a part of a chain, does the parent have a track record one way or the other elsewhere in the country? CI can help them answer these questions and plan for their future.

Similarly for attorneys, there is more to running a legal practice than merely practicing law. For example, the world of the very large law firms has been churning. Large firms have merged and become larger. Other firms have lost entire practices to competitors. Some have ceased being "one stop shops" or spun off boutique practices. Still other long-established firms have vanished—in one case over a single weekend.

- In every case, there were impacts on every other law firm from the largest to the sole practitioner. What kinds of impacts?
- Spinning off a boutique firm could create a new competitor for you in your own specialized practice, perhaps even locally.
- Law firms merging and dividing throw off, unfortunately, a lot of highly trained, somewhat shocked, attorneys. Many will be looking for work; other to set up their own practices. Where will these practices go and what will they focus on?
- When law firms merge, some clients will not make the journey due to conflict of interest issues. Can you identify some of them before they "come on the market"?

A smaller local firm suddenly becoming a part of a larger regional or even national firm may soon find some existing clients or practice areas less desirable. Do you know what they might be and how to benefit from that?

In each of these cases, a little CI can help prepare not just the sole practitioners and small firms, but also the largest multi-national firms, to be proactive and benefit while others just look on—surprised.

References

Anon (1987) Getting to the heart of Japanese R&D. High Technol : 7–9 (February)

AnythingResearch (2011) Market size matters. http://www.anythingresearch.com/industry/Tax-Preparation-Services.htm. Accessed 1 Sept 2011

Association for Strategic Planning (2011) Application—the richard goodman strategic planning award 2012. http://www.strategyplus.org/pdf/AwardApplication_2012.pdf. Accessed 11 Oct 2011

Burgess, Congressman Michael, Optional one page tax form. http://burgess.house.gov/flattax/. Accessed 1 Sept 2011

Carpe D (2003) Diggin' deep online to profile an executive–Part 1. SCIP 1:32 (Online)

Gieskes H (2001) Competitive intelligence at Lexis–Nexis. In: Prescott JE, Miller SH (eds) Proven strategies in competitive intelligence: lessons from the trenches. Wiley, New York

Global Intelligence Alliance (1/2006) Does your business radar work? Early warning/opportunity systems for intelligence (White Paper)

Global Intelligence Alliance (1/2008) MI for the strategic planning process—case examples (White Paper)

Hovis JH (2001) CI at avnet: a bottom line impact. In: Prescott JE, Miller SH (eds) Proven strategies in competitive intelligence: lessons from the trenches. Wiley, New York

Jaskiewicz SP (2011) Trade secrets: do they exist in era of web, social networks? East Pa Bus J 11 (January 31)

Kaltenheuser S (2002) Working the crowd, Across the Board (July/August)

Lavelle L (2001) The case of the corporate spy: in a recession, competitive intelligence can pay off big, BusinessWeek Online. http://www.businessweek.com/magazine/content/01_48/b3759083.htm Accecssed 11 Oct 2011 (Nov. 26)

McKinsey & Co (2008) How companies respond to competitors: a McKinsey global survey. http://www.mckinseyquarterly.com/How_companies_respond_to_competitors_2146. Accessed 11 Oct 2011

Miller J (2000) Small business intelligence–people make it happen. In: Miller J (ed) Millennium intelligence–understanding and conducting competitive intelligence in the digital age, medford. Cyberage Books, NJ

National Institute of Standards and Technology (2011) 2011–2012 Criteria for performance excellence. US Department of Commerce, Washington

Roop J (1998) Low cost lowdown. Inside Bus

Sedulo Group (2011) Potential acquisition. http://www.sedulogroup.com/case_studies/potential_acquisition.php. Accessed 4 Oct 2011

Society of Competitive Intelligence Professionals (2004) CI effectiveness in support of marketing. www.SCIP.org. Accessed 12 July 2004

WestBlog (2008) Using IP for competitive intelligence. http://tnalcorpcomm.wordpress.com/2008/10/28/using-ip-for-competitive-intelligence/. Accessed 11 Oct 2011 (October 28)

Chapter 4
Preparing Yourself

4.1 Get Your Own CI Toolbox

Over the past 40 years or so, Competitive Intelligence (CI) specialists have accumulated a significant toolbox of analytical tools and techniques. Some of these, such as SWOT analysis, have been adopted from other business disciples, such as strategy development. CI practitioners also refined others, such as patent mapping. Still others, such as psychological profiling, were developed by the behavioral scientists and adapted to CI. Later in the book, we have provided you with a list of them and additional information on some of them.

4.2 You and the CI Professionals

After deciding that doing at least some CI research and analysis yourself is a good idea, make sure you understand the relationship between you and the rest of the CI community. Most people who talk about competitive intelligence are talking about it to other people who also all do it on a full-time basis. That is not the case with you. For you, CI is one of many tools in your management skills toolkit; for these others, it is a full-time discipline or even a profession.

When you dip into CI's literature (see Chap. 13), think of yourself as if you are in the medical profession. There, we have many specialists who have developed their own literature and skills, and generally communicate only with other specialists. Then, we have the general practitioner, or family physician: someone who is expected not only to know all of the basics of many different disciplines, but also must know when and how to hand a patient off to a specialist, and when that is not appropriate. You are like that general practitioner. That is, you have to know enough about CI to conduct it yourself for your firm, but you must also understand your limits in conducting competitive intelligence and when you need to bring in a "specialist".

J. J. McGonagle and C. M. Vella, *Proactive Intelligence*,
DOI: 10.1007/978-1-4471-2742-0_4, © Springer-Verlag London 2012

For example, it would be virtually impossible in many situations for you to pick up the telephone and call your opposite number at a competitor firm. While you would feel comfortable with calling its customer service number, or other support center, which by the way, can be good sources of information, it's unlikely you would feel comfortable with directly calling a product manager at a competitor. In fact, some firms have specific rules in their codes of ethics and conduct that would bar you from doing this. The reason they have these restrictions is related to antitrust compliance: stopping employees from being put in situations where it can be claimed later on that they were conspiring to fix prices or divide markets. So at that point, if you want calls to be made to a competitor, you have to bring in a "specialist".

Keeping this in mind you have to understand that the "specialist" also has an advantage that you do not possess. That is, he/she comes to your problem with a broader perspective. Or to put it another way, he/she does not suffer from the intellectual blinders that can constrain you.

4.3   Your Intellectual Blinders

Noting that you have intellectual blinder is not meant to be critical of you. Rather, it reflects the fact that all of us have developed certain patterns that govern how we do and see things. We all have developed certain sets of preconceptions that are based on our past knowledge and past experiences. The problem for all of us is that we do not review and update these every single day. That means we are caught, from an intellectual point of view, in a rut.

What kind of a rut? Actually it is ruts—plural. Here is a partial list of the kinds of bias and misperceptions that you could fall victim to:

- Best-Case Analysis: An optimistic assessment based on your intellectual pre-disposition and general beliefs of how others are likely to behave.
- Denial of Rationality: Attribution of irrationality to others who your perceive are acting outside the bounds of your own standards of behavior or decision making.
- Mirror-Imaging: Perceiving others as you perceive yourself.
- Presumption of Unitary Action by Organizations: Your perception that the behavior of others is more planned, centralized, and coordinated than it really is.
- Presumption that Support for One Hypothesis Disconfirms Others: Evidence that is consistent with your own preexisting beliefs is allowed to disconfirm other views.
- Rational-Actor Hypothesis: Your assumption that others will act in a "rational" manner based on your own rational reference.
- Wishful Thinking (Pollyanna Complex): Excessive optimism born of smugness and overconfidence.
- Worst-Case Analysis (Cassandra Complex): Excessive skepticism. (Adapted from NATO 2001, pp. 46, 47)

How can you get yourself out of that rut?

The first step is accepting the fact that this is true—for you. No, this does not mean that you are impaired in some way. Just recognize that everyone, no matter how smart or how objective or how skilled, operates in the same way. Knowing that this is the situation is the first step leading you to the ability to recognize these facts and preconceptions.

Second, you have to understand and identify what is the "conventional wisdom" within your firm, your own discipline, and your industry. In many cases, conventional wisdom is just that, wisdom. In many cases, however, it is merely conventional, but hardly wisdom.

Let us illustrate that with a real-world example:

- Several years ago, we worked indirectly for firm that was long-established in its market niche. It dealt with a particular commodity that had a limited number of uses. As a condition of receiving government assistance, it had to bring in a completely independent, outside firm to review certain elements of its business plan. We were retained to review the competitive part.
- Without getting into too much detail, we found that the firm, which had been in this business for over 25 years, had much right, but several critical things were wrong. By running the firm on the basis that "We have been doing this for 25 years and we know what we're doing", their blinders had locked them in. They had never looked outside their industry to see if there were other uses for the product whose manufacture they dominated. A quick search of patents found that there were, in fact, new, emerging uses for this product. That could have been very good news for them as producers.
- On the other hand, a search of the medical literature indicated that there were studies underway, but not yet completed, that might link this particular product to severe health consequences for those who worked with it downstream. That meant that this firm was facing potential future regulations on its end-use and perhaps even on how it was manufactured.
- Why did the firm miss these? It missed them because it was looking every day in the same places it had looked for 25 years. There had been no new uses for this product in years, so it stopped looking for them. That did not mean the new uses were not being developed, just that it did not see them. The same is true for the medical or more correctly, the potential medical, problems.

So how do you get yourself past this? Do not necessarily stop using the information sources that you have now. If there is industry publication that you read every week or every month, continue to read it, just with a slightly more skeptical eye. If you receive newsletters, make sure you review them. However, begin to read them, play this particular game: what is the *real* source of the story? That is, did the newsletter get the story from a press release? And did the magazine get the story from the newsletter? You will be surprised how often you will see this happening.

Beyond this, you must learn to broaden your horizons, and to do this on a consistent basis. One very good way to do this is to read things that other people do not read, or listen to sources or visit websites that other people do not use. We do not mean merely about your industry, but across the board.

For example, if your firm is involved in running restaurants, in addition to reviewing restaurant magazines, why not subscribe to *Information Week* magazine, which will look at your industry from a very different perspective? If everyone in your office always reads the *Wall Street Journal* every day, perhaps you should read the *Financial Times*, *USA Today*, or the *Chicago Tribune*. If everyone in your office reads *Time* magazine or *Business Week*, perhaps you should be subscribing to the *Economist*.

Now, by reading these, we do not mean cover to cover every time, although that is occasionally a very good exercise to widen your perspectives. You can subscribe, and use a smart phone, iPad, or laptop computer to scan these new publications. The goal is for you to begin looking at things from a different perspective, so to get you started looking in different places.

The same should be true in your private life. What do you read for recreation: mysteries, biographies, comic books, or medieval history? Try something different. If you read mysteries, pick up a good biography or history book. It does not matter what it is about; what matters is that you are training your mind to look beyond those intellectual blinders that we all wear. If your hobbies do not include reading, but gardening, then bring a portable radio out and listen to something on the radio, say, a sporting event or music, or even talk radio. The idea is you are always looking, you are always listening, and you are always learning.

Going beyond this, learn about CI, more than we can teach you in this book. Read books about competitive intelligence, attend meetings, or conferences that have CI as one element and make sure you attend that session. For more on that, check out Chap. 13.

When you go to that trade show or industry conference that you have gone to every year for the past 5 years, take a longer look at the list of all the seminars, workshops, and other events open to you. Are there groups or subjects that you have traditionally avoided? If so, make it a point to attend one of those sessions. Realistically, you may find out that your decision not to attend in the past was well-placed. But by going to a financially-oriented presentation when your area of focus is using social media for marketing, you open yourself up to meeting different people, to having to master a slightly different discipline, and to receiving new ideas.

Try it. It works.

How do you know that it is working? You will not have a road to Damascus moment. Rather you will find that you read even the same resources differently. If you read a local newspaper every morning, but never read the lifestyle or sports sections, you may find yourself one day reading about the latest negotiations of some professional sports association, or the newest suggestions on how to "live cheap". We are not saying you will gain an insight from one of these pieces that will change your direction, but you will continue to teach yourself how to learn. And that is the most important way to keep your mind open and to be able to conduct your own CI proactive research. Remember, unlike the CI "specialist" who has a separate client, and perhaps even separate researchers, you are the client, the analyst, and the researcher. Keeping your mind open continuously is your way of adjusting for the fact that there are no other people who are checking your work in a competitive intelligence context.

Reference

North Atlantic Treaty Organization (2001) NATO open source Intelligence handbook. http:// www.au.af.mil/au/awc/awcgate/nato/osint_hdbk.pdf. Accessed 11 Oct 2011 (November)

Chapter 5
Figuring Out What You Really Need to Know

5.1 How Can You Get the Right Kind of Competitive Intelligence to Help You in Your Decision-Making?

Remember, your goal is develop Proactive Competitive Intelligence—that is CI that enables you to act, not just react. So, what do you really need?

- Early warning of possible competitive threats? How can you identify them if they have not happened? Start by looking at what has happened to your firm in the past 2–3 years. What surprised you? Which of those surprises could you have spotted if you had been looking for it? Is there something that most or all of them have in common?
- Strategy development help? Begin by determining what assumptions you, or others, make about what your competitors will do. Then what do you assume they will do in response to your planned actions? Were you right about those assumptions in the past?
- Sales? What do you really know about competitor pricing and marketing? Probably just what your customers tell you—and they are not always a reliable source. How do competitors set prices? Have recent changes in marketing surprised you? If you knew more, what would you do with it?

Remember, the key is to identify what would be Proactive CI for a specific target. *Always ask yourself what is the end use*: that is, what would you do differently if you had that CI right now? If the answer is clear, great—you have a CI target. If it is fuzzy, you are into the "need to know versus want to know" dilemma. Proactive CI means to you need to know, just not want to know, because your knowledge will drive a decision or an action.

Until you have been doing this for a while, you will find that you need some help in really defining your needs. Without clearly defining your needs (which are not necessarily the same as the questions you have), you cannot hope to do

J. J. McGonagle and C. M. Vella, *Proactive Intelligence*,
DOI: 10.1007/978-1-4471-2742-0_5, © Springer-Verlag London 2012

first-rate research for raw data, nor can you hope the analyze it to extract everything of value from it.

Let us take a quick look at some basic principles for you to keep in mind when doing your own competitive intelligence. Keep these in mind as you work your way through learning *how* to do this:

- *Admit that you have not always been focused on what your competitors are doing*: Even if you have been trying to keep up with what the competition is doing, your efforts are almost certainly sporadic and incomplete. If you are not really keeping up, you are probably just assuming you know what the competition is doing. Never assume you know what your competitor is doing, and, more importantly, never assume you know what it is *planning* to do!
- *Know who your real competitors are*: They may not be who you think they are. Ask your customers what other firms else they considered before they chose you. Those should be considered competitors, too. And keep an eye on your partners, suppliers and even major customers. They could quickly turn into competitors.
- *Familiarize yourself with the competition—as they really are—today*: Take the time to visit and revisit their stores, study their plants (if possible), check out their web sites, and find out who owns them. Look for information about competitors in the public domain—press releases, newsletters, government filings, etc.
- *When you study your competitors, never assume they see things the way you do*: Your competitors have their own vision of the marketplace—and of your firm. Even if you think that vision is dead wrong, always keep in mind that they are guided by it and will operate in accord with it.
- *Decide what's important—and what is not*: There are some things you cannot do anything about no matter how much you know about them. Focus on supporting decision-making, not satisfying your curiosity. To be Proactive, get only the data you need for important decision-making, make sense of it—and then act on it!
- *Do not assume there is nothing you can do, even if you know what your competitors are up to*: Effective CI does not always provide an opportunity to develop a competitive advantage, such as launching a new product. Sometimes, it provides an early warning that helps you survive!
- *Ask lots of questions*: If a customer leaves, find out why he/she is leaving and where he/she is going. Keep track of the answers you get. You may see a pattern that warns you of new competitors or new initiatives. Now you know what to focus on.
- *Do not get pressured into trying to measure exactly what CI is doing*: While there are many aspects of CI where you can measure the impact, you cannot attach a number to everything CI can do for you. (McGonagle and Vella 2002). For example, what is the value of knowing a competitor will beat you to market or knowing that a competitor's newest initiative will run into problems because the construction of the plant supporting it still lacks some key permits?
- *Be realistic*: With increasing security on all fronts, some sources of raw data CI that were available in the past are no longer open to the public. Others may not be in the future. Always keep these changes in mind. (For more on this, see Chap. 13).

- *Do it right—or do not do it at all*: CI is an ethical, legal activity. Never let yourself get pressured into doing anything that is not totally ethical and legal. There is never any good reason to be unethical or illegal.

5.2 How do You Plan Your Own CI Research?

While every executive and manager, regardless of the firm's size, its product/service, or its competitive environment, should be doing some sort of CI, not every firm can afford to do all types. However, how do you figure out what it is that you personally really need to know?

We are going to offer you two proven approaches: the use of a checklist and a "blank screen" process. Feel free to use any or all of these, together, separately, or in combination. The key is to help you to determine what Proactive intelligence you need to do your job better.

5.2.1 Get Off to a Good Start

There are five issues involved in conducting your own competitive analysis:

- First—what are you looking for and why?
- Second—what limits are there on your research?
- Third—where might the information that you need, in CI, usually called the raw data, be located?
- Fourth—having at least initially determined where that data may be located, how you get to it?
- Fifth—once you have that data, how will you make sense of it so that you can use it?

Keep in mind that these same issues arise when you ask others to provide CI to you. We will deal with that aspect a little later in Chap. 10.

5.2.1.1 What are You Looking For and Why?

Answering this question can be more difficult than it actually seems. Most people try to collect everything there is out there, which is an endless and ultimately pointless effort. You have to step back and redefine/refine what you are looking for—before you start.

What specific intelligence are you looking for? A better way to put it is, if today you had the intelligence that you are looking for, what decision would you be able to make now that you cannot make without it? In other words, is what you are seeking Proactive?

If you cannot answer this question "yes", then your task is not refined enough. You risk ending up with general information, rather than Proactive CI.

Next, you should ask yourself, why do I want this intelligence?

- Is it to better understand the competitor? If so, the amount of effort you should dedicate to this should be fairly small—you are now in, or least, very near, the world of "nice to know", not the world of "need to know".
- If it is to help conduct a "business war game", which we deal with in Chap. 11, then your research has to be broader and more in depth. You have to provide you and your team with original material, such as advertisements, biographies and the text of interviews, rather than mere summaries of secondary business magazine articles.
- If you are trying to figure out your competitors' intentions and capabilities, then you will have to be the looking not only at bricks and mortars, capital and cash flow, but at people as well. Why? Because people make decisions; buildings and invested capital do not. They merely reflect decisions already made.

Before going any further, you should look around, literally and metaphorically. Ask yourself, "Do I already have some of this data? If I do, should I use it? That is, is it both reliable and current?" If so, you have at least a starting point. If the data is not current, it still may at least provide you with an indication of where you might seek the updated data today, as well as potential people to talk to. If it is not reliable, treat it as such.

5.2.2 What Do You Need?

5.2.2.1 Using Checklists

A pre-designed competitor analysis checklist can help you expand or refine your CI needs, simply by forcing you to identify both the specific targets and the relevant data you will initially seek. That is, are you seeking data on specific direct (or indirect) competitors? Which ones? Are you seeking data on designated lines of business? What lines? What types of data are you seeking—micro-level or macro-level? Are you looking at history, the present, or the future?

Using a checklist is most useful when you are trying to develop CI on a number of competitors, as it assures that you will try to collect the same data on all of them. When done this way, it is also useful in running a competitor analysis, as described in Chap. 11.

We are giving you with a sample checklist. Using a checklist like that below provides one important way to make sure you are not omitting anything critical when considering the scope of any research you are involved in. Use this as a template to develop a checklist customized for your own needs and your specific competitive environment.

Sample Checklist for Competitive Intelligence Research

Instructions: Prepare a separate checklist for each competitor, or project, even though the industry-wide information may not change from competitor to competitor (unless the competitors are in different fields or countries). Place a check to the left of each item that you have identified as needing research. You should not check too many. If you do, you either are working on a very large project, or you have to find another way to narrow its scope.

Overall Industry and Competitive Environment

☐ Industry or industry niche involved
☐ Current competitive environment
    ☐ Industry structure
    ☐ Number of competitors, their product lines (or range of services), and locations
    ☐ Market shares, gross sales, and net profitability of competitors
    ☐ Expansion potentialities of competitors
    ☐ Important differences among competitors
    ☐ Industry marketing, distribution, and pricing practices
    ☐ Rate of technological change in this industry niche
    ☐ Need for new technology
☐ Barriers to entry and exit
☐ Regulatory constraints and political pressures
☐ National/global economic, scientific, etc. events that can impact the competitive environment
☐ Potential entrants/future competitors
☐ Attitude toward likely new competition
☐ Indirect competition

Individual Competitor Information

☐ Identification
    ☐ Full name
    ☐ Name/acronym commonly used
    ☐ Ultimate parent
    ☐ Ownership history
☐ Key owners and managers
☐ Major shareholders/partners
☐ Directors and officers: backgrounds and other business relationships
☐ Corporate and management organization: formal or informal
    ☐ How decisions are made
    ☐ Management styles, abilities, and emphases

☐ Depth, capabilities, and weaknesses of management in key functional areas
☐ New personnel and recent restructuring
☐ Corporate politics
☐ Products/services offered
    ☐ Current offerings
    ☐ Current and potential future applications of products/services
    ☐ History of key products/services
    ☐ Forthcoming products/services
    ☐ Products/services likely to be changed, or eliminated
    ☐ Quality of product/service
☐ Customer service policies and performance
☐ Distribution channels, including strengths and weaknesses
    ☐ Possible changes in distribution channels
☐ Financial and legal positions
    ☐ Short- and long-term borrowing capacities and ability to raise equity financing
    ☐ Sources of financing and nature of relationships to sources
    ☐ Sales margin, return on assets, and return on equity
    ☐ Profitability of key divisions, products, and services
    ☐ Projections of financial position over next two to five years
    ☐ Comparison of profitability, cash flow, and other key ratios with those of major competitors
    ☐ Significant liabilities
    ☐ Major lawsuits and regulatory actions: probable impacts on competitor
☐ Personnel, resources, and facilities
    ☐ Labor force: cost, availability, turnover, and quality
    ☐ Union status, contracts
    ☐ Sustainability, productivity, environmental, and diversity programs
    ☐ Joint ventures, minority interests, and other investments or ownership interests in facilities
☐ Supply chain
    ☐ Supply chain management structure and strategies
    ☐ Quality control programs in place or planned
    ☐ Manufacturing and operating costs
    ☐ Facilities: locations, current performance, and potential
    ☐ Planned improvements to existing facilities or new facilities
    ☐ Planned facility closings or divestitures
☐ Make/buy/partnering policies
    ☐ Outsourced activities
☐ Sales force and customers
    ☐ Type of sales force
    ☐ Organization of sales force
    ☐ Training, capability, and compensation of sales force
    ☐ Number of customers
    ☐ Distribution and concentration of customers

- ☐ Analysis of largest or most important customers
- ☐ Targeted customer base
☐ Sales and pricing
- ☐ Commercial, nonprofit, and government sales
- ☐ Domestic versus foreign sales
- ☐ Seasonal/cyclical issues
- ☐ Pricing strategy: who prices products/services and how
- ☐ Credit, discounts, incentives, consignments, and other special pricing policies
☐ Marketing
- ☐ Market shares by product line, by geographic area, and by industry segment
- ☐ Marketing approaches and current effectiveness
- ☐ Marketing and service capabilities and reliability
- ☐ Samples of advertising, product literature, and promotional materials
- ☐ Samples of products/services
- ☐ Probable future changes in marketing direction and timing
- ☐ History of questionable marketing practices
☐ Technology, research, and development
- ☐ Current manufacturing methods/processes
- ☐ Key patents and proprietary technology
- ☐ Access to, use of, and dependence on outside R&D and technology
- ☐ Need for new technology
- ☐ Pending changes in manufacturing methods and processes
- ☐ Size and capabilities of research staff
- ☐ Usual lead time between R&D breakthrough and delivery of a product/service to market
- ☐ Types and levels of R&D, including current and likely future expenditures
☐ Competitor's strategies, objectives, and self-perception
- ☐ Business philosophy
- ☐ Corporate strategy
- ☐ How strategy is made and executed
- ☐ Targeted markets and market shares
- ☐ Target financial objectives
- ☐ Technological objectives
- ☐ IT strategy
- ☐ Recent improvement/restructuring initiatives and their results
- ☐ How the firm sees itself
- ☐ How does the firm see its competitors and its customers

5.2.2.2 The Blank Screen Process

Another option for developing your own CI research strategy is most useful when you are starting only with a general sense of what you need. Here, you start with a blank screen on your computer opened to word processing. Then fill it with what you think a final report might look like. To do this, you ask and then answer, as best you can now, the following:

- What points are critical for you to answer, dealing with the subject matter of your question(s)?
- What raw data do you probably need to make a clear statement or draw a conclusion for each of those points?
- How specific and current does that raw data have to be to allow you to draw a conclusion and to answer the question(s)?

This approach starts you at the very end of the project, dealing with the goals you have, rather than at the beginning. This way you can minimize being influenced by preconceptions about what you may expect to find and where. Instead, you set as your goal the raw data you think need and then work backwards to identify exactly what data you need and then go and locate that specific data. This entails four basic steps.

Step 1. To develop your research goals, outline the first page of a possible report on your project. Make sure that outline shows a possible recommended course of action. Begin by restating the needs statement as a conclusion and as an action statement. For example, if the question is "How can we catch up with competitor A in terms of costs of production?" you can restate it as follows: "Our competitor's most significant cost advantage is in its use of nanotechnology systems in its manufacturing activities. [Note: Remember, this is only hypothetical.] For our firm to catch up, we must change our manufacturing processes to incorporate the newest Asian nanotechnology".

Step 2. Following this action statement, list the kinds of key findings that might tend to support this recommendation. Following this example, you could write: "Our competitor has, and will continue to have, substantially lower costs than we do for manufacturing the same products. This is due to its nanotechnology plus access to lower cost inputs. The experience of manufacturers in other, similar industries indicates that nanotechnology can be successfully adapted to our manufacturing operations within a short period of time, providing important cost savings."

Step 3. Now, again transform these statements into questions. Convert the conclusion you drew in Step 1 into the following:
- "Does our competitor actually have lower input costs?"
- "Why are these costs lower?"
- "For the two most important reasons for lower costs, can we duplicate or improve on these?"

You are now drawing closer to a statement of the focus for your own CI research. Knowing the research focus is crucial. It enables you to control and limit your data-gathering to those that have the greatest likelihood of helping you deal with this problem.

Also, once you know the focus of the research for the project, you will also be more likely to identify other data of potential interest, or at least leads to that data. For example, here, if your research disclosed that a major consulting firm had just created a nanotechnology consulting practice, you might want to retain that information for future reference.

Step 4. Now, repeat the conversion process in Step 3 for all of the draft key findings you developed in Step 2, like this: "Does our competition have substantially lower input costs than we do? If they do, will that continue? Is existing nanotechnology sufficiently developed to be able to support our manufacturing techniques? What has been the experience of manufacturing firms in other, similar industries with the same or similar nanotechnology? Specifically, can it be successfully adapted to our manufacturing operations within a short period of time, and will this provide a cost savings?"

If you need to, you can repeat the process until you have created a series of even more focused questions on which you can focus your CI research efforts. As you progress, you should find that, if any of your small assumptions are not correct, you can quickly adjust your focus, by changing the questions, while the research is underway.

5.3 Data-Gathering Limitations

The direction your research will take will reflect the nature of each project. But, no matter what the project, you will always have to deal with these limits:

- Time
- Money
- Self-imposed constraints
- External constraints

5.3.1 Time

Time is usually the most critical limit on your CI research. There are a wide variety of time constraints, some or even all of which may be applicable to a particular project. They can be most easily understood by asking, and then answering, the following questions—honestly:

- How many hours of your time will your data collection probably take?
- How much of your time is available for your analysis?

- How much time do you have to conduct the entire project? That is, when can you start and when do you have to make a decision based on what you find?
- How long will it take to collect the raw data that you *really need*, that is, to answer the most important questions, but not necessarily all questions? Remember, it is often better to have a pretty good answer on time than a perfect one which is 2 weeks too late.
- Can you start your analysis work with only partial data? If so, when? Or with what data?
- How much time do you really have available, and when? That is a function of your schedule, and exactly how important intelligence that you could produce is to you.

Often, the best way to proceed is to determine when you need the CI and then work backwards. For instance, if you have four weeks to develop data on the market plans for new products of two key competitors, planning to go to a trade show or industry conference that will be held in 2 or 3 months is obviously not a viable option.

5.3.2 Money

Financial constraints control more CI gathering decisions than most people are willing to acknowledge. That is, how much money do you have to spend, how much can you do alone, how much help can you get from others within your organization at no cost, and do you have an outside CI research firm you can turn to for assistance?

For example, should you hire a CI consultant who plans to interview key executives in person or should he or she do this over the telephone? In person interviews are likely to produce more useful raw data, but will cost considerably more than using the telephone. And using the consultant will cost more than if you do it—although you may be limited to whom you can talk with.

Time and money are always trade-offs in CI research. For example, your research strategy may identify three possible avenues of research. If you have sufficient time, you might logically pursue the one with the highest likelihood of success first. Then, if that does not provide enough CI, you would move to the second, and so on. However, if you are under severe time constraints, you might have to engage in parallel activity. That is, instead of proceeding sequentially, you may have to start all three at once. As you do not have the time to do otherwise, the total costs must rise accordingly. As a rough rule of thumb, cutting the time available for a project by half once it is ready to start or is underway can be at least double the costs of acquiring the same raw data. If time constraints become more onerous, this can cause costs to escalate even more.

5.3.3 Self-Imposed Constraints

These constraints can vary widely. While you may be comfortable with contacting your competitors directly, you may find this is barred by a policy of your firm. This may force you to approach projects from a different angle.

Another constraint may be a legitimate concern about allowing too many of your co-workers learn about your CI. For example, you may not want other employees to be involved with the assignment at all, as in the case of preliminary research involving a potential acquisition.

5.3.4 External Constraints

These constraints are more intangible. They arise from your firm's written and unwritten policies, as well as from legal and ethical considerations.

You should always check your firm's policies on data collection, as some, for reasons unrelated to competitive intelligence, may prevent you from directly calling a competitor. This is usually due to historical concerns about antitrust issues. Some firms may have a narrower prohibition, such as telling their employees that they may not discuss pricing issues with any competitor. This is obviously to avoid price fixing. Whatever the limits are, you owe it to yourself and your employer to be aware of them and to abide by them.

The legal limits on competitive intelligence, to the extent that any really exist, deal with how information is collected. Thus, you cannot not steal materials from a competitor, misrepresent for whom you work, pretend to be a student doing a paper, or any of a number of other dubious, if not illegal, steps. Note that most of these actually can fall within the issue of ethical behavior, not legal behavior. For convenience, we have placed a longer discussion of these legal and ethical issues in Chap. 6.

The only legal regime directly applicable to CI is the US Economic Espionage Act of 1996. That act deals with the theft of trade secrets. If you come into possession of a trade secret of a competitor, the safest thing to do is to immediately contact your firm's attorneys, bring them all the materials that you may have received accidentally or otherwise, and let them handle it from there.

Some firms that have a CI unit have, or should at least have, a set of standards governing the unit's performance. So if your firm does have a CI unit, even though you are doing work on your own, consider yourself to be bound by whatever those rules are. For an idea of the range of issues that surround the ethical collection of competitive intelligence (McGonagle and Vella 2003), you may want to look at the Code of Ethics issued by the Strategic and Competitive Intelligence Professionals in the Appendix to Chap. 6.

References

McGonagle JJ, Vella CM (2002) Bottom line competitive intelligence. Quorum Books, Westport
McGonagle JJ, Vella CM (2003) The manager's guide to competitive intelligence. Praeger Publishers, Westport

Chapter 6
Managing Your Own Research

6.1 Effective CI Research Planning

There are several ways that CI analysts approach potential data sources when designing their data gathering. We will cover several of their proven approaches for you in the next chapter. Here we first want to help you understand what is involved in doing your own research.

6.2 Research Guidelines

Be realistic but optimistic. If you can determine what you need, the data is almost always somewhere to be found. To locate where the raw data you need for your CI may be accessible, as we detail in this chapter and the next, remember to think about whom or what else has an interest in that data and why they have an interest. Each person and/or organization you can identify as having an interest is a potential source for your raw data, as well as a potential linkage (also described later) to other data sources. For example, if you are interested in the sales of in certain departments in retail clothing stores, using these criteria could lead you to trade associations, corporate annual reports, and trade publications.

Do not limit yourself with, and do not be limited by, indexing or what turns up first in your online searches. For example, if you have identified a trade publication of potential interest, check whether that publisher has a special annual issue, or even a separate publication that covers topics of particular concern to you.

J. J. McGonagle and C. M. Vella, *Proactive Intelligence*,   53
DOI: 10.1007/978-1-4471-2742-0_6, © Springer-Verlag London 2012

6.2.1 How Likely are You to Get What You Need?

Always think about the likelihood that you will get good data from each target source you have identified. Let us assume that you have concluded that a particular industry association may be a good source for some of the raw data you need. Now think: what is the likelihood that the association will have exactly what you want, that it will provide it to you (if you or your firm is not a member), and that this all can happen within your time and cost limitations? Similarly, if your project involves manually reviewing the back issues of a local newspaper for want ads placed by a target competitor two and three years ago, is it really likely that you can access a complete set of issues for those years, and that you will be able to review them all in a cost-effective manner? If not, before you start on such an effort, you should try to identify alternative approaches or alternative sources of data that may satisfy your data needs.

6.2.2 What Do You Need First?

Throughout your research, maintain a clear idea of the relative importance of the data you are seeking and of the data sources you are trying to locate. Do not waste time tracking down minor pieces of data to produce a picture-perfect product. CI is not academic research, but rather a commercial activity, one in which the final product often ages rapidly and badly.

As you develop more experience, you may see that your progress typically depends on collecting certain kinds of data first. For example, if you are going to interview key personnel at a competitor's distributors, you must first identify the distributors, and then identify the key people at each one before you can start interviews.

Also, some data will arrive more slowly than other data, even when you have located where it is. To understand the trade-offs involved in timeliness, compare the following potential ways of obtaining the contents of documents filed by a target competitor with a municipal zoning board in another state:

- You could ask the target firm for copies, but it is extremely unlikely to respond positively. Remember, it does not have to send you any of these materials. Also, asking the target firm for these documents will probably alert it to your interest—not a good idea.
- You can see if the local zoning board has a web site where files they have are indexed or even may be accessible. If they are accessible online, you are done. If not, you will have then to contact the local agency and arrange for it to provide you with copies. This may not happen very quickly.
- If the local agency only allows people to review the files in its office, you could find and then hire a local commercial service to copy the documents directly from the files, scan them, and email them to you or air express or fax them directly to you. This is more costly, but faster than relying on the zoning board.

6.2.3 Tiering Your Research

Sometimes you may not be able to find the exact data you think you need, quickly, cheaply, or easily. If a project looks like it could turn out to be excessively demanding, consider dividing your work into tiers, or stages, of research. Then design your research to take you from one stage to the next stage, while, at the same time, narrowing the number of targets or scope of the research for the next stage.

For example, let us say you need to determine the US capital expansion plans of ten competitors. Start by taking all ten firms and first determine whether or not each is planning any significant capital expansions. If three of the ten are not planning any significant expansions, you can drop these from the research. Then focus on getting detailed information on the remaining 7 targets.

What you are doing is attacking the most general level first and going after the most easily obtained data, which you can review quickly. Next, you analyze that data in terms of what it tells you about what additional, more specific data, you need to complete your task, as well as where that data may be located.

Then, move to the next tier or stage. At that point, you continue to narrow the scope of the research, perhaps by eliminating potential targets or data sets, and then collect more data and analyze these results. If possible, you drop additional targets in the third tier, perhaps because their planned investments are non-US, continuing in this way until you are done. By using this process, you can often save both time and money, because you focus your resources on locating the most critical bits of more specific data in the second and following tiers.

6.2.4 Linkages

When you look for raw data, you should never mentally restrict your search. Not only can one source provide several different and important types of data, but multiple sources should be cross-checked to generate data on the same point. In addition, you can use one source to lead you to another source. In an article, for example, see who is interviewed, even if they did not comment on your any of your concerns. They may have something to tell you that did not make it into the article. Or they may be able to connect you to someone who can help you.

6.2.5 Secondary Before Primary

Going about getting the data requires that you have at least some idea where it might be located. Generally, you should start collecting secondary data before moving onto primary data, that is, interviews and the like. The secondary data will usually provide you with a historical look at the subject/target. In addition, you should review it for the names of individuals and organizations that you may want to contact for further information. This generates leads to people who may be experts quoted in an article, an executive of a competitor who is no longer there, a supplier, or even a regulator.

We have some more to say on primary data collection in Chap. 8.

6.2.6 Be Creative: Try New Things

If your research takes you onto the Internet, you will almost certainly turn to one of the major search engines. Let us assume that is Google, but it could be any other.

When you first run your search, you will undoubtedly come up with thousands of hits. Before wading into these list, scan them to look for subjects that are clearly not of interest to you that are included. Then, instead of narrowing your search on this search engine, go to another one, say Yahoo, and run your new search there.

You will be rewarded with a lot of hits, but they will be more focused. When reviewing them, can you reformulate and narrow your search further into a short question. If so, try that question on Bing.

What you are doing is exploiting the sweep of the search engines and also their strengths. By shifting from one search engine to another, you are forcing yourself to do more to refine your search than you would do by just going to a more advanced search on the same engine. What happens when you do that is often you lose track of what 3, 4, or 5 refinements you have made, so you cannot replicate that search in the future. When you shift search engines, you have to note what your search strategy is, so you can keep track of it.

6.2.7 What Should Your Outlook Be?

In the classic CI situation, there is interaction between the analyst/collector and the end-user in terms of defining needs, the competitive context, expected outcomes, and the end-use of the intelligence to be provided. There may also be interaction between an analyst and a separate data collector. None of that is applicable in your situation. You do not have interplay with a second person.

Such interplay, even with someone who is working in the same firm, provides a way of avoiding or least minimizing fully internalized biases. To put it simply, the mere act of questioning and answering between two people will help avoid this. Of course, as time passes, it can be harder in the classic CI situation to do this as the collector/analyst and end-user work together more and more.

A healthy approach for you is to consider looking at radically different theses to test them. That is, accept nothing as given even if it has been established in a previous research project. That forces you to revisit your own assumptions and separate out assumptions and biases from established facts.

An alternative way is to mix up your questions, that is ask different questions and approach the questions from different angles. For example, conducting your own version of environmental scanning is probably a healthy concept. That scanning forces you to look at the environment first, but not to determine what the answers are to questions that you have; rather, you should be trying to find out what is really happening "out there" now and therefore what is it important to know more about.

For you, the collector/analyst who is also an end-user, the problem is difficult because there is no outside person with whom you can engage in discussions to help test assumptions, to check findings, or to dig out long-held biases that impact the way you view the world. This requires an increased degree of discipline on your part. That discipline may be supported by simple aids such as actually writing down what it is you are seeking to learn and what you think you know, emphasizing the word think as opposed to know.

The real goal here is to identify and maintain a state of intellectual rigor at all times. This does not mean you need to create new tools, but rather, to inject a careful and cautious attitude toward any long-held beliefs or personal knowledge to avoid completely contaminating your research.

There are also analytical techniques can be used to help avoid falling into these traps.

- In reviewing the data that you collected, it is sometimes useful to arrange it in ways other than the way will be analyzed at the end. For example, if the data is to be supporting a particular research project which shows a chronological development, it may be useful for you to randomize the dated materials so that you are read them out of chronological order. The point is to force you to read what is there, and not read what you think is there.
- Another technique is to avoid a stopping too early. The most common problem will be that you will stop when your pre-existing thesis or belief has been confirmed. If you come to the end of the research and find that your thesis has been proved, it would be useful to do the research for a little bit longer to make sure that, in fact, you have not overlooked data contradicting that key thesis.

6.3 Just How Much Data Do You Really Need?

When you start digging, you may find that your research is going very quickly, particularly if you have planned it out and identified your best sources at the beginning (see Chap. 7 for more on this). In that case, you run the risk of being buried in data. Resist the temptation to "dig just a little deeper". When you have your answer, and you have confirmed that it is correct, go onto something else. The additional refinement you may get from additional research may not be worth it.

Also, again, before you start, figure out if you are looking for current awareness, that is, what is going on now, as opposed to going on a "deep dive". Do not forget you can always go back for additional research, but you cannot get back the time you loose with unneeded work. If this is possible, keep notes on what you found, and where, as well as where you might go if you had additional time. That way, if you go back, your research path is already partially laid out for you.

Working by yourself, as opposed to being part of an existing competitive intelligence team, requires that you develop a different perspective from that CI professionals would have. This is not to say that you are already approaching the problem of analysis incorrectly, but is just recognizing the fact that, as we noted earlier, we all work with blinders on, at least metaphorically.

Recent research indicates that when you affirmatively remove yourself from the problem you are dealing with, that is mentally stepping away from the problem, you actually do see it differently. For example, let us say that you are faced with the problem of a new product coming into the marketplace from a well-positioned, but smaller, competitor that goes head-to-head with you in an adjacent market space. We could be talking here about food, lawn service, electronics, etc.—it does not matter.

In this situation, let us also assume that a member of your sales team has told you that she spotted this new product on the shelf in a store that carries your products. Now, your first reaction as a product manager might be "what's going on here?" If you begin your research here, you are certainly going in the right direction, but you are not focused enough yet, and you may not get the right data or undertake the right analysis to be proactive.

Stand back and look at the problem again. Put yourself in the position of someone who does not understand why this small competitor is so important to you. Of course, it is important that someone is launching a new product against you, but why is *this* firm so important? Now, in our hypothetical, you have run into this firm before. You found it to be a scrappy competitor, disciplined in its approach to the market, and very difficult to dislodge once it becomes entrenched. So for you, the real issue is not the product; it is the sudden presence in your market space of a proven, difficult competitor.

What does this mean for your research? It means that you should be asking yourself, not only why is it doing this, but how is it doing this? As we indicated the firm is a smaller competitor. Would not a third-party ask you, "What have they done to be able to bring an additional product to the market? That is, are they in a new joint venture, or just labeling someone else's product? Have they shifted over their production to the new product, which could mean producing less of the established product? Have they built a new facility or purchased an existing facility to make this new product?"

You now have a very different perspective, because you are focusing more on the competitor, and not so much on the product. If we just dealt with the question "What is going on here?" the answer might be, "They are introducing a new product which is a complement to the existing product". From there you should have to ask additional questions that drive additional research such as "What are they expect to do with it?", "Where is it being made?", "How is it being distributed?", and "Do they intend to take this national?"

By standing back and articulating the underlying assumptions, underlying concerns, or underlying knowledge, you can immediately begin to focus more directly on your problem, on your research, and on what answers you really need, from a different perspective. This is not to say that you can be completely objective and completely shake off the blinders of past experience, corporate preconceptions, and the like. But since you have no one that is helping you with this, you must put yourself in two places at once, as it were.

If you are somewhat put off by the idea of stepping away or developing a new perspective, try this trick: write down what it is that is bothering you, in this case, finding a new product from an unexpected, but potentially troublesome source. Do not just write down finding a product, and the other implied conditions. Then write a two sentence explanation of why this is an important subject to look into, pretending, perhaps, you are asking someone to help you with this and need to give them a sense of direction. Then, turn your statements into questions. Now you already have a more refined approach.

If the standing back or trying to adopt a different perspective does not provide assistance at the beginning, do not worry. It may well be that what you are dealing with is something that is so vague that it is merely unsettling. Go with your gut.

But after some initial research, stop, and try to redefine your problem in light of what you have learned so far. Think of it as breaking into stages. At the end of each stage, you should ask yourself "Has my question or questions changed? Have some of my questions already been so well answered so that no further work is necessary? Have additional questions emerged which need to be dealt with now?"

Here are several tips from experienced CI researchers about managing the course of their research:

- Make sure you at least tried to access some resource in each of the large resource groups that we have given you in later chapters, such as Government or Media. It is not that each group will always have the data you are looking for, but without at least trying to dig into each one, you cannot be sure that a particularly useful source of data has not escaped you.
- Stop about halfway through your research. Look at, either physically or in your mind, where you have gone. Have you found that interviews have been much more useful than secondary research? Has what you been looking for been disclosed, at least in part, by filings with federal, state or local governments? Use these insights to reset the balance of your research.

6.4 Finishing Your Research

In planning and executing your research, be thorough, but not obsessive. In CI, you must learn to accept something that is less than perfect, but still accomplishes the desired end. In other words, many times you have to say to yourself, "It is close enough?".

How do you know when you are done? There is an informal test we have found very useful. It is called "tail chasing" or, more elegantly, "closing the loop". This occurs when new sources of raw data seem continually to give you leads that take you back to sources your previously exploited, or at least identified. At that point, you are usually at the end of your basic research.

6.5 Planning for Future Research

If you face the possibility that you will have to conduct research against any one of an identifiable set of targets in the future, but you do not know when that will happen again, it might be useful to develop a written summary of the potential target, and, more importantly, where you can locate critical data on it if and when you need it again. Do not rely on your memory for this. By spending a small amount of time now determining where data on your target will probably be when you eventually need it, when you go against this target again, you will already have an idea of where data might be, and just as importantly, where it is not likely to be.

6.6 Conducting Supplemental Data Gathering

A logical follow-up to a first round of data evaluation and analysis is to identify any supplemental data you need to collect. Typically, the kind of data you will need at this point is narrow and focused, enabling you to answer new, highly focused questions that have arisen as you have done your research and analysis. But just because the data you need is supplemental, do not ignore the processes outlined throughout this book. Since you may find that the narrow, highly focused questions can be harder to answer than were the broader questions you started with, it is all the more important to approach this in a disciplined manner.

One caution: If you need some supplemental data, by all means seek it out. But do *not* use this as an excuse to redo your earlier collection efforts.

6.7 Establishing and Maintaining Your Own Network

In the field of governmental intelligence, there is a saying that goes something like this: 80–90% of all the intelligence you need is in the public domain (Marrin 2011, p. 108). When looking at your business, it can safely be said that 80–90% of the intelligence that is accessible to you in the public domain probably lies somewhere in your own firm.

Information is an odd commodity. While it loses its value quickly, people that possess information never really let go of all of it. While economists have studied this phenomenon, we can look at it very simply: if you read a sheet of paper and you give someone else that sheet of paper, you may have given them the document, but you have retained part of it yourself—in your memory. It is that retention factor that is important to understand when you do your own CI.

Within your firm, there are individuals who know bits of information about your competitor, but do not understand or appreciate how valuable the bits are. Since you cannot contact everyone and then order them to tell you everything they know, not that you would want to do that, you have to find some other way to access the tremendous potential of the data held within your firm.

This data really is made up of three types: information in information systems, documents that people have obtained and retained for their own purposes, and the knowledge that people carry in their heads. Regardless of where that data is, for you to get to it, you have to go to a person. That means that you must work affirmatively to develop an internal network of individuals who can help you when you need it.

6.7.1 How to Go About That?

Think of yourself as a salesperson. Every salesperson uses some kind of contact manager to keep track of the individuals whom they have met and how to get back in contact with them again in order to make a sale, or to get their help in making a sale. Visualize your own networking the same way.

You should go through your current contacts within your firm, (the same rules apply for sources outside of your firm), and indicate on your contact manager in some way when the last time it was that you contacted each person and for what kind of data they might be a useful resource in the future. Then contact them again. Tell them that you're trying to build your own capabilities to do your job better by adding CI as a tool, and ask them if they will assist you. Make it clear that you are not asking them to do anything in addition to their existing job; you are just asking them to be available to you should you have a question that someone with their expertise, and do not be afraid to call it expertise, can help you with. That expertise may be knowledge of clients, of suppliers, of distribution channels, or of the news media. But they possess that expertise and in addition, they probably possess their own networks, formal or informal.

Every time you go to a firm meeting, make sure you know every person that is there; if possible, get a list and add them to your contact manager with a simple note that you met them at the quarterly briefing in March 2012, for example. Then, if you need to contact them again, you can remind them that you have met and under what circumstances. If you have not met, but you were both at the same meeting, at least you have a way to open a conversation with a person.

You must never abuse these contacts. Use them carefully. Do not call people at random and hope that you may get lucky. Have a specific reason for calling specific people, and explain that you do not want to waste their time, but you were looking to see if they could give you a little help or could suggest somebody who could help you. If they can suggest someone, thank them, get the full contact information, and as with the interviewing process (see Chap. 8), ask whether or not you can use their name in introducing yourself to the new contact.

For your network to truly work, you must make several things happen:

- People must at least casually know you so that they can be comfortable talking to you.
- Individuals that you talk to must be made to feel secure that what they say will not come back to haunt them. If you are going to eventually distribute a report to someone, you should ask each individual "Can I use your name in the report as the source for this information" before you begin the conversation. If he/she seems the slightest bit uncomfortable, then tell him/her that you will simply say, "A member of the Southwest field force has indicated that..."
- After a particular project is over, it is a nice touch to send an e-mail or leave a voice message saying "Thanks for your help, it was really important". You want people to understand that their assistance is appreciated and that it was useful.
- You must always be willing to help the members of your network as quickly as you would have them help you. And, you should tell them that. What you want to engender is a reciprocal relationship with gratitude on both sides. In other words, each party in this relationship should feel that there is a current, past or future benefit to staying in that relationship.
- Depending on the size of your network, it is probably a good idea to find a reason to contact all of the people in it from time to time. By that, we do not mean weekly or monthly or even quarterly contacts. Perhaps every six months or so, make a brief call or e-mail people on your contact list to see how they are doing, and to keep track of job changes and promotions. In particular, if you find that someone who has been useful to you in the past is now in a new position, look at this as an opportunity to double your opportunities. First, that individual is still part of your network, but it is up to you to find out what he/she is doing now and to convert that into an opportunity to help you develop future competitive intelligence. In addition, if the help that this individual has provided in the past was dependent on the position that he/she held, asking for a brief introduction to the individual that replaced your contact is a way to keep your network vital and growing. Then contact that new person.

6.8 Legal and Ethical Issues

Since you are working by yourself, you do not have the usual checks and balances that you might associate with another task done on a team basis. That means you have to be sensitive to legal and ethical issues and constraints.

The biggest problem you face is that, in many people's minds, possibly even in your own, CI is associated with spying. Let us get that right out of the way. Spying or, more correctly espionage, is a crime. You are not engaged in criminal activity. If you are, you do not need this book and we do not want you saying you involved in CI in any way.

Above all, never operate in a way that makes you feel uncomfortable. This is the first and best level of protection: your own personal set of ethical and moral standards. If you are uncomfortable with an approach to research, or with talking to particular people because of their current position or past employment, then do not do it. Many people suggest that a personal code of ethics may well be easier to define if we put it in a different context: *Never do anything that you would want to see reported the next day on the front page of your local newspaper.* So that, if you should determine that "dumpster diving" is in fact legal in a particular case, do you still really want to do it? If not, then do not do it.

What happens if something, say a document, comes your way that you are not expecting, something that is clearly confidential or worse? Have an action plan in place first, so you do not have to decide on the fly what to do. The bottom line is that it is improper, in some cases illegal, for you to obtain information that you know are a competitor's trade secrets. However, remember that the trade secret laws discussed earlier do require that the person who claims something is a trade secret has a legal obligation to take significant steps to protect it. To put it another way, just because someone puts a rubber stamp on a document that says "trade secret", does not make that document a trade secret, if the individual then hands out several hundred copies of the document at a tradeshow.

Tradeshows themselves pose some additional issues, which we cover in Chap. 8. In our experience, going to a tradeshow and simply listening carefully, can be extraordinarily valuable. Even though people know who you are and for whom you work, because of your name tag, once they feel comfortable, they are often going to talk to a third-party about things that may interest you in your presence. Do not be afraid to listen.

The question arises from time to time, "I cannot get to talk to anybody at a tradeshow because my name tag has my firm's name on it. So what can I do about it?" The answer is, not much. If you are in a corporation where there are numerous subsidiaries, *and* you actually work for several subsidiaries at once, it is possible that you could register for the meeting under one of those lesser known names. But remember, we said that you must actually work for it, as opposed to just assuming such a bland employer name. It is totally inappropriate for you to go to a tradeshow and use a fictitious personal name and/or a fictitious business name on a tag. It is also unethical for you to replace any kind of coded badge, which designates exhibitors from advertisers, media, etc. with one which disguises the kind of business for which you work. There is plenty of information to be gleaned from a tradeshow without crossing these lines. See Chap. 8 for more on trade shows and industry conferences.

The same thing is true in telephone interviews. You are never "a student" calling for help on "a paper". If you think that is approach is appropriate, then you should not be doing this kind of work. In addition to being unethical, the ironic fact is that an individual who poses as someone else tends to reduce his/her odds of obtaining the kind of raw data that he/she actually seeks and needs. If you are "student", then the data you may get will be data appropriate for "student". For more on conducting telephone interviews, see Chap. 8.

You should also be very sensitive to your firm's policies, both those that are written, and those that are unwritten. Review the written ones, carefully. You are not just looking for rules on conducting competitive intelligence research, for these are few and far between. Rather look at those rules that govern your conduct or contact with individuals outside of your firm. You may find that it is considered improper for you to contact someone directly at another competitor. Obey that. You may find that you are not permitted to trade in certain stocks, or trade any stocks, without clearance from legal counsel. You should inquire whether the reasons underlying this would also restrict your ability to contact people from those firms. Again, think ahead.

The unwritten rules at your business are the most important. What underlies most of them is one word—embarrassment. Do not do something that could cause concern to your employer or bring unwanted attention to it. Whether or not there is a written policy that says this, the cold facts are that taking such action puts your job at immediate risk. In that regard, let us tell you a story, a short one.

Several years ago one of the largest consumer goods firms in the US, which had a well-regarded competitive intelligence unit, authorized a research project against a global competitor. The details are not precisely clear, but it appears that the first CI firm with which the client contracted then brought in a second group of firms as subcontractors, and some of these subcontractors in turn subcontracted some of their work to other firms. That meant that some individuals working directly on the assignment were at least three levels away from the client's supervision. The results were predictably catastrophic: One of the subcontractors was accused, by the target company, of attempting to steal its garbage in order to go through it later.

Events then moved rather quickly. The CEO of the client firm flew across the world to meet with the CEO of the target firm. The end result was as follows:

- The client firm paid the target a substantial settlement, believed to be over US$10 million;
- The client firm also agreed with the target that it would not enter a certain market niche for a period of years, the very niche that was the focus of the competitive intelligence task;
- On the client firm's CEO's return to headquarters, three or four competitive intelligence personnel were quickly terminated;
- A senior CI manager at that firm "retired" rather quickly; and
- The client firm purged its approved contractor list, removing every firm that was involved in this case, even the firm which claimed that it blew the whistle on the misdeeds of others.

The client firm paid a rather substantial price a failure of its management and a failure by others to exercise common sense. Keep in mind the words of a CI pioneer, Professor Stevan Dedijer:

Intelligence today is about using the collective knowledge of the organization to reach an advantageous position in industry. Spying is dying—only idiots resort to these kinds of shady activities. Only companies with an inadequate intelligence capability and with inferior knowledge-acquisition strategies seek to obtain information by illegal or unethical means (Bloom 2004).

6.9 Appendix: Ethical Codes

Excerpts from the Strategic and Competitive Intelligence Professionals "Code of Ethics for CI Professionals":

***To comply with all applicable laws, domestic and international.

To accurately disclose all relevant information, including one's identity and organization, prior to all interviews.

To avoid conflicts of interest in fulfilling one's duties.

To provide honest and realistic recommendations and conclusions in the execution of one's duties....

To faithfully adhere to and abide by one's company policies, objectives and guidelines (Strategic and Competitive Intelligence Professionals, 2011).

Excerpts from The Helicon Group's Ethical Policies:
The Helicon Group:

Collect and disseminate data in full compliance with applicable state and national laws....

Never employ questionable data collection activities. These are techniques, otherwise legal, which, if made public, might tend to embarrass Helicon's reputation or that of a client.

Supply honest and realistic evaluations of what data and analysis can and cannot be developed for a client before beginning an assignment.

Accurately disclose all relevant information, including the caller's identity and organization, prior to all data collection interviews.

Respect all requests for the confidential handling of information provided during data collection....

Comply with all applicable client policies and guidelines dealing with the collection of data for competitive intelligence.... (The Helicon Group, 2011).

References

Bloom D (2004) Stevan dedijer. http://www.guardian.co.uk/news/2004/sep/01/guardian obituaries.obituaries. Accessed 29 Sept 2011, Aug 11 2004

Marrin S (2011) Improving intelligence analysis: bridging the gap between scholarship and practice. Routledge, New York

Strategic and Competitive Intelligence Professionals (2011) Code of ethics for CI professionals. http://scip.org/About/content.cfm?ItemNumber=578&navItemNumber=504. Accessed 20 Sept 2011

The Helicon Group (2011) Ethical policies. http://helicongroup.com/Ethics.htm. Accessed 20 Sept 2011

Chapter 7
Identifying the Best Sources of Raw Data

7.1 Visualize Where the Data Might be Located

Before starting your research for data, sit back for a moment and consider where the data may be located. Data, or as economists call it, information, is a "sticky" commodity. This means that someone who has been in possession of some piece of data still has a part of it, even when it has been passed it on. That portion may be in the form of a copy of a document, of a report which integrates some of the data, or just in the individual's own memory.

One way to start your research is to determine where data like that you need, but on your own firm, is located. Having found where data on your firm is in the public domain, or at least is accessible to diligent research, you are better prepared to locate similar data on your competitors.

All of this leads to determining where is the most likely place that the exact data you need is located. Do not be limited at this point by considerations of whether or not you have the ability to access that source. Rather, you are trying to understand from where it originates, so that you can then try to determine, or at least estimate, how it moves from the originating source to others. These other locations may be within the competitor or outside the firm. They may be employees, former employees, consultants, customers, suppliers, catalogs, surveys, etc.

Your goal here is to understand how the data moves, who changes, consolidates, or divides that data, and then determine if and how you can identify and approach each of these sources. You may find that the data you are looking for is not available from any one source, but has been divided up among several. This is typical of pricing information. This means you will have to approach several sources, and (re)aggregate the data that you collect. Only when you have first made an effort to understand the origination and flow of data, should you begin to seek it.

Listing the potential resources also allows you to control your research. We are often asked how we know when we have run out sources for data. Our answer is for you to keep a written inventory of the places/people/institutions approached or even identified as sources for the data, even if you have not approached them.

J. J. McGonagle and C. M. Vella, *Proactive Intelligence*,
DOI: 10.1007/978-1-4471-2742-0_7, © Springer-Verlag London 2012

When you find that as your research continues, new sources are referring you back to old sources, it usually means your research is done. That is closing the loop.

7.1.1 Viewing Competitor Data as a Commodity

Instead of somehow scanning the universe of potential data sources at the beginning of your research, you could approach the task by thinking of the raw data on your competitor that you are seeking as a commodity–a tangible product. When you do that, then ask and answer the following questions:

- Who produces the raw data I want?
- Who collects the raw data I want?
- Where is the raw data that I want transferred, and why?
- Who else uses the raw data I want?
- Who accumulates the raw data that I want?
- Who else has an interest in the raw data I want?

In answering these questions, you will quickly narrow down where you can begin to look for the raw data you need to develop the CI you need.

7.1.1.1 Producers of Raw Data

For example, say you are seeking some firm-level or even divisional-level data. This would seem to mean that the targeted competitor actually produces it—somewhere. But, you should dig further. Specifically, who at the competitor is most likely to produce this raw data? For example, is R&D strategy determined by the engineering department, dictated by a strategic plan prepared by the planning department, or the result of input from each? If the answer is that the R&D strategy is a blend of engineering and planning department inputs, when you are, for example, interviewing personnel formerly with a targeted firm, make sure you get people from the appropriate department involved in the process.

7.1.1.2 Data Collectors

The person or department you identify here may not be the same person or department that actually produced the data. The key here is transmission: when data is being transmitted, it is first assembled and sometimes analyzed, but always moved. One key to locating raw data is to determine where the data is moving so you can try to intercept it, in a figurative sense only. That is why the next issue also becomes important.

7.1.1.3 Data Transfers

Assume that for any number of reasons, your firm wants detailed but apparently unavailable data on the performance of the US gambling operations of a casino. Perhaps you want to determine whether the casino is so profitable that your competitor can expand, or thinks that it is losing so much money that the competitor will have to delay expansion plans in the other areas where it competes head to head with you.

At least check with the Casino Control Commission for the state or states in which the competitor has casinos. These commissions have a number of areas of interest in that same subject, one of which may be the profitability, or at least dollar volume, of the casino. And because these are public agencies, the data the commissions require casinos to provide to them may be considered as nonproprietary and can therefore be made available to you.

7.1.1.4 Data Users

To determine who uses the kind of raw data you want to locate, look for individuals, such as securities analysts, who want data from firms to generate their firm and/or industry forecasts. For you, what the analyst has to offer is first of all, his or her analysis and conclusions. However, of potentially greater importance may be the raw data—whether numeric or narrative—provided by the target competitor to the analyst. That data may, in turn, be disclosed in the analyst's reports. If not, you may be able to get it by direct contact with the analyst. If not, you may still be able to extract it from the reports by carefully reviewing them.

7.1.1.5 Data Accumulators

Every source that accumulates data will have data of differing types and quality. The type and quality of that data depends on why each of them collected the data in the first place. For example, organizations such as the U.S. Census Bureau and trade associations generally (but not always) work with aggregated data, as opposed to data at the firm or lower level. But, using disaggregation may enable you to isolate significant data that you and the providers of the data both assume to be masked by the aggregation efforts of those providing the data.

7.1.1.6 Others with Interests in the Data

To learn what other organizations and people may have already collected some or all of the data you are seeking, your focus will quickly move to a relatively wide range of potential resources. These can range from the advertising departments of industry publications to academic research centers. Industry publications' advertising departments may collect and generate data similar to what you are seeking to show that their publication represents the focused audience an advertiser wants to reach. To do that, the advertising department may collect data, or even commission special surveys, to educate potential advertisers about the industry and about its key participants.

Academic centers that focus on a particular industry can be a useful resource because their access to data is sometimes much freer and/or broader than that of trade association. In addition, the firms that they deal with will sometimes assume that giving academic researchers access to the data is "harmless", in a competitive sense. Of course, some competitors are savvy enough to insist that the academics who are given access to competitively sensitive data do not release any of the data, except in some aggregated form.

7.1.1.7 Internal Resources

Do not forget to consider looking within your own firm. Start with your network, as we discussed in Chap. 6.

Also, ask human resources if there are people who formerly worked at a competitor who now work for you. When approaching them, make it very clear that there is no obligation for them to talk about their work at their former employer. Make sure they understand that no one is putting any pressure on them. What you want is voluntary cooperation. If they feel constrained for ethical, or even legal, reasons, then end the conversation.

What you should do in any interview or attempted interview, whether internal or external, is to get another name that you can contact, preferably using the name of the person you spoke to as a reference. Opening the door that way is extremely valuable to develop useful primary information.

We will deal in greater detail with approaching your co-employees for help with your CI work a little later.

7.2 Scoring to Identify Good Sources of Raw Data

We have developed a system to help CI professionals determine which one or more of four different types of intelligence are most likely to be the intelligence processes most useful for them (McGonagle and Vella 2002, pp 89–114). Here, we will give you an abbreviated version of it. This system focuses on letting you identify what sources are likely to provide you with the data you need.

To do this, we ask you to consider five elements:

- Your competitive environment,
- What products/services you produce,
- The firms you face,

- Key production and supply chain issues, and
- The dynamics of your marketplace.

Read the descriptions that follow and decide what description best coincides with your competitive environment. You will then use this to fill out a score sheet following this section. While doing this takes time, unless your competitive world and competitors change radically, you should only have to do this once.

7.2.1 Your Competitive Environment

Here you should look at three areas that make up your competitive environment:

- Level of Company Regulation
- Barriers to Entry
- Barriers to Exit

7.2.1.1 Level of Company Regulation

This means regulation that may be aimed at the competitor because of the industry in which it operates, such as auto insurance, or it may be regulation which is aimed at the competitor because of how it operates, such as all manufacturers that discharge products into the air. Or it may be aimed at the firm because of how it is structured. For example, a firm which is publicly held and which trades on the New York Stock Exchange faces very different regulatory restrictions from a privately held competitor.

Companies facing the lowest levels of firm regulation are, in general, freest to act and to respond to market forces. In addition, as they have low levels of regulation, they tend to be able to innovate more quickly, to change their operations and even overall structure more rapidly, and thus tend to place less information in the public domain.

Companies that face a moderate degree of firm regulation are less free to act and to respond rapidly to market forces. They take longer to be able to respond because regulation directly or indirectly impedes their flexibility. Also, they tend to place more information in the public domain than do the firms facing the lowest levels of firm regulation.

Companies that face a high degree of firm regulation are usually least free to act and to respond to market forces. In addition, as they face the highest levels of regulation, they tend to innovate relatively slowly, and are unable to change their operations and even overall structure as rapidly than firms facing less regulation. Finally, they tend to place the greatest amount of information in the public domain.

7.2.1.2 Barriers to Entry

Barriers to entry are numerous, varied, and sometimes inter-related. Some are purely regulatory in nature, that is a firm must obtain a license, which entails getting the approval of a government agency, before it can open a bank branch. They may be financial, such as the high cost of building a new plant, or they may be technologically based, such as the possession of defensive patents by firms in that industry.

Companies in a market that has low barriers to entry typically are characterized by higher degrees of flexibility and innovation than are those in other markets. Because of the absence or minimal nature of these barriers, such markets are subject to relatively sudden changes in terms of new participants as well as new products.

Companies in a market that has moderate barriers to entry usually are characterized by lesser degrees of flexibility and innovation than are those in markets with low entry barriers. Because of these barriers, such markets are not subject to sudden changes in terms of new participants as well as new products.

Companies in a market that has high barriers to entry are characterized by significantly lesser degrees of flexibility and innovation than are those in markets with low entry barriers. Because of these high barriers, such markets are not subject to sudden changes in terms of new participants as well as new products. Instead, such changes are much more gradual and often incremental in nature.

7.2.1.3 Barriers to Exit

Barriers to exit are not found as often as are barriers to entry, but they do exist. For example, a financial and regulatory barrier to exiting the long-term care insurance business might be the requirement that the insurance firm establish reserves to cover all potential future claims which might be made based on past policies, and that a licensing body has approved the adequacy of such reserves. Another barrier to exit might be a requirement imposed on a private emergency medical service firm that it may not surrender its ambulance operating permit on less than a 180-day notice. A third type of barrier to exit is the existence of a massive specialized capital base, such as a hospital center, which cannot easily, quickly or inexpensively be converted to other uses.

Companies in a market that has low barriers to exit typically are characterized by higher degrees of flexibility and innovation than are those in other markets. Because of the absence or minimal nature of these barriers, such markets are subject to relatively sudden changes, in terms of participants leaving as well as products being taken off the market.

Companies in a market that has moderate barriers to exit are usually characterized by lesser degrees of flexibility and innovation than are those in markets with low exit barriers. Because of these barriers, such markets are not subject to sudden changes in terms of withdrawing participants as well as withdrawn products.

Companies in a market that has high barriers to exit are characterized by significantly lesser degrees of flexibility and innovation than are those in markets with low exit barriers. Because of these high barriers, such markets are not subject to sudden changes in terms of departing participants as well as products taken off the market.

7.2.2 Products/Services Produced by Your Competitors

Here, you are looking at the specific products (or services) which your competitors provide to the marketplace in which you all are now competing. This is subdivided into four separate areas:

- Number of direct substitutes
- Level of Regulation of the Product/Service
- Time to Market
- Life Cycle of the Product/Service.

7.2.2.1 Number of Direct Substitutes

What is a direct substitute for your product/service should be viewed from several perspectives:

- Your perspective, that is what products/services do you see, in the market today, serving as a direct substitute for your product/service?
- Your competitors' perspectives. That is, do they see your product/service as a substitute for their products/services? If they do not think so, determine why not. If they do, what other products/services do they appear to consider as direct substitutes for their own? Are they also direct substitutes for yours as well?
- What do those consumers, or other customers, to whom you are seeking to sell your product/service, see as direct substitutes? Their perspective may surprise you.

Companies whose product(s)/service(s) face very few direct substitutes also tend to face only a few directly competing organizations are often protected from competition because of patents or proprietary processes, and are in a marketplace that may have significant barriers to entry, to exit, or both. They tend to be involved in more "head to head" competition, with a greater emphasis on product/ service differentiation than on price, and with some vulnerability to sudden impacts due to technological changes outside of their marketplace.

Companies whose product(s)/service(s) face a moderate number of direct substitutes are typically characterized by the fact that they face several directly competing organizations are poorly protected from competition by patents or proprietary processes, and operate in a marketplace that may have only moderate barriers to entry, to exit, or both. Because of these factors, they tend to be involved in more "head to head" competition with competitors as well as product line competition, with less emphasis on product differentiation and more on price, and with real vulnerability to sudden impacts due to technological changes outside of their marketplace.

Companies whose product(s)/service(s) face a large number of direct substitutes are typically characterized by the fact that they also face a large number of directly competing organizations, that they are rarely able to rely on any long-term protection from competition by patents or proprietary processes, and operate in a marketplace that may have virtually no barriers to entry, to exit, or both. Because of these factors, they tend to be involved largely in "head to head" competition with competitors, product-to-product competition, and product line competition, with very little emphasis on product differentiation and most of the focus on price competition, and with regular impacts due to technological changes outside of their marketplace.

7.2.2.2 Level of Regulation of Product/Service Produced

The level of regulation of the product/service produced cannot always be separated easily from regulation of the competing firm, particularly when dealing with services. However, it is useful to try to make that distinction.

Regulation of a product/service can cover some or all of these areas:

- Price at which it can be offered, such as for milk producers;
- Terms under which it can be offered, or which it must include, such as individual auto insurance policies;
- Quality, that is minimum standards, of the product/service, such as in the case of automobiles in the US where each auto maker must produce a fleet of cars with a certain overall minimum gas mileage;
- Restrictions on who can serve as an intermediary, such as rules governing the transportation of human organs for transplants; and
- Limitations on to whom or where a product/service can be sold, as in the case of prescription drugs.

Companies that are operating in an environment with little or no regulation of the production and sale of their products/services are typically characterized by the fact that they face a market that changes rapidly. In addition, often there are multiple, flexible channels of distribution, so pricing decisions can be quickly transmitted to the marketplace, and competition is on the basis of perceived differences in product and service characteristics and consumer value.

Companies that are operating in an environment with moderate levels of regulation of the production and sale of their products/services are typically characterized by the fact that they face a market that changes less quickly. While existing products/services can be modified or replaced, that is not done often or quickly. In addition, often there are fewer, less flexible channels of distribution, so pricing decisions are transmitted to the marketplace less quickly. Competition is typically conducted on the basis of differences in product/service characteristics as well as on brand-name recognition.

Companies that are operating in an environment with high levels of regulation of the production and sale of their products/services are typically characterized by the fact that they face a market that changes slowly. While new products/services can be introduced, their introduction is not easy, so that they do not appear with great frequency. Similarly, while existing products/services can be modified, that is not done very often or very quickly. In addition, often there are very few, relatively inflexible channels of distribution. Competition is typically conducted on the basis of differences in product and service characteristics as well as on brand-name recognition.

7.2.2.3 Time to Market

Time to market is, as with so many other concepts, relative. In some industries, such as the pharmaceutical industry, the time from the beginning of research and development until a product can be taken to the retail market can be measured in decades. In contrast, in the software industry, it may take only months to convert an idea to a marketable product.

Companies facing an environment where there is a long lead-time to market typically see the marketplace occupied by relative few players. These players are concerned with long-range plans, and positioning their product/service portfolios to assure a continuing and growing stream of revenues over time. In addition, such markets are often, but not always, dependent on using existing proprietary technology to make the production and/or distribution more efficient and thus squeeze out additional costs.

Companies facing an environment where there is a moderate lead-time to market typically see the marketplace occupied by a number of players of varying sizes. These players are concerned with positioning their product/service portfolios not only to assure a continuing and growing stream of revenues over time, but also are alert to respond to shorter-term changes. In addition, such markets are often developing their own technology to make the production and/or distribution more efficient and thus squeeze out additional costs.

Companies facing an environment where there is a very short lead-time to market typically see the marketplace occupied by a large number of players of widely varying sizes. These players are less concerned with positioning their product/service portfolios to assure a continuing and growing stream of revenues over time, than to assure that they identify and then respond to very short-term changes. In addition, such markets are often using their own technology, as well as technology from other sectors, to continually force the production and/or distribution to be more efficient and thus squeeze out additional costs.

7.2.2.4 Product/Service Life Cycle

The product/service life cycle can be thought of as the time it takes the product/
service to move from being new and revolutionary, through becoming established
and evolutionary, to becoming passé and a commodity. As the product/service
moves through these stages, its unit profitably tends to fall, it sales volume tends to
rise and then fall, the number and variety of competitors as well as competitive
products/services tends to rise, and the number and variety of products which
incorporate it in some way first rise and then fall.

Companies that offer a product/service that has a very short life cycle face a market
that changes very rapidly, one where prices and costs can take sudden and significant
turns. Typically the supporting investment is not relatively significant, but the role of
both propriety and non-propriety technology is. In addition, the products/services
usually cannot be protected by legal regimes such as patent and trademark filings.

Companies that offer a product/service that has a moderate life cycle are in a
market that changes quickly, but where prices and costs rarely take sudden and
significant turns. Typically the supporting investment is somewhat significant, as is
the role of both propriety and non-propriety technology. In addition, the products/
services usually cannot be completely protected by patent and trademark filings.

Companies that offer a product/service that has a long life cycle face a market that
rarely changes quickly, that is prices and costs rarely take sudden and significant
turns. Typically the supporting investment is significant, but the role of both propriety and non-propriety technology is much less so. In addition, the products/ser-
vices usually receive significant protection from patent and trademark filings.

7.2.3 The Firms Your Firm Faces

Here, you are evaluating the firms with which you directly compete in your market
or markets. The ways these firms are managed and relate to each other have a
direct impact on where you can find the data you need. This is, in turn, divided into
five separate areas of consideration:

- Centrality of management;
- Level of cooperation;
- Degree of concentration;
- Level of knowledge intensity; and
- Reliance on alliances.

7.2.3.1 Centrality of Management

Here, you are taking into consideration how the firms you face tend to be managed.
While each firm differs from every other firm, experience shows that firms in the
same markets often tend to have similar degrees of management centrality.

By centrality of management, we mean the extent to which critical decisions, affecting the product/service and market, are made at the highest levels of an enterprise, by central managers, or, conversely, are made at the level closest to the unit that is actually producing, selling, or otherwise managing the product/service. By decision-making, we mean that the final decision is made at one level, whatever level that is, and is not subject to review at another level before it is implemented.

By critical decisions, we mean decisions such as these:

- Changing prices and characteristics of existing products/services;
- Developing new products/services;
- Launching new products/services and removing existing products/services from the marketplace;
- Authorizing capital investment needed to support these operations; and
- Entering into long-term contractual relationships with other firms necessary to carry all of these steps out.

In markets where your competitors tend to be centrally managed, you should expect that their response time to market and consumer changes will take longer. In addition, decisions, such as those listed above, will be made only after taking into account the needs and demands of other, non-related operations, as well as the need to satisfy a variety of potentially conflicting constituencies, such as employees, unions, governments, owners, and investors.

In markets where your competitors tend not to be totally centrally managed, but where some decisions are made at lower, more local levels, you should expect that response time to market and consumer changes will be faster. Decisions, such as those listed above, often will not only take into account the needs and demands of other, non-related operations, as well as the need to satisfy a variety of potentially conflicting constituencies, such as employees, unions, governments, owners, and investors, but will also reflect local pressures, such as those generated by consumers and local interests. over a moderately long time.

In markets where your competitors tend to have the most decentralized management, you should expect that response time to market and consumer changes will be quite fast. Decisions, such as those listed above, may occasionally take into account the needs and demands of other, non-related operations, as well as the need to satisfy a variety of potentially conflicting constituencies, such as employees, unions, governments, owners, and investors. However, they will more often reflect local pressures, such as those generated by consumers and local interests.

7.2.3.2 Degree of Cooperation

By degree of cooperation, we mean the degree to which the firms in this market engage in vigorous, rapid moving competition. Those markets in which the firms seem to have the least degree of competition may have high levels of cooperation, or even cartel-type behavior. On the other hand, markets with the lowest levels of cooperation will be ones that would commonly be described as being highly competitive, that is, each firm operates without any evident parallelism in conduct.

In markets where the firms are highly competitive, they have the lowest levels of cooperation. Such markets tend to be fast moving, and the actions of one firm cannot often be a useful predictor of the actions of other firms.

In markets where the firms are more cooperative, the overall degree of competition is somewhat less. In these markets, for example, some firms may engage in joint research, or joint marketing efforts. Others may actually use competitors as a source of product/service for this market. Such markets tend to be slower moving.

In markets where the firms appear to be cooperating to the greatest degree, they can be said to be operating like a cartel. In those circumstances, the effective levels of competition are even lower. In these markets, for example, most of the firms may engage in joint support of research, or of product development. Such markets tend move much more slowly. Given high degrees of cooperation, at least tacitly, usually the actions of one firm can be a useful predictor of the actions of most of the other firms.

7.2.3.3 Degree of Concentration

A similar situation involves an evaluation of the degree of concentration. But while the previous test deals with conduct, this one deals with structure. In general, the more concentrated is a market, the less intensive and extensive will be direct competition. The ultimate degree of concentration is the monopoly, where one firm controls 100% of a market: there is no competition there.

We created the following scale, based on the tests used by antitrust regulators, which measures sales, either in terms of volumes or in value:

- A market is "highly concentrated" if the 5 largest firms control more than 60% of the market;
- A market is "moderately concentrated" if the 5 largest firms control at least 25% but no more than 60% of the market; and
- A market is "slightly concentrated" if the 5 largest firms control less than 25% of the market.

In a highly concentrated market, the largest firms tend to dominate the competitive landscape. In general, they tend to set and defend prices, and also indirectly control the prices of key inputs, simply because of their (relative) size. It is relatively difficult for firms outside of the top 5 in size to penetrate that tier or to significantly increase market share. The competition tends to be at two levels: among the very largest, and among the rest, with little real competition between either group.

In a moderately concentrated market, the largest firms tend to heavily influence the competitive landscape. In general, while they tend to set and defend prices, they are vulnerable to competitive initiatives from smaller, more aggressive firms.

Also, the very largest firms have less influence over the prices of key inputs. While it is difficult for firms outside of the top 5 in size to penetrate that tier, it is not impossible. The firms outside of the top 5 can significantly increase market share even without penetrating the top 5. The competition tends to be among the very largest, and then among all firms in the market.

In a slightly concentrated market, the largest firms tend to have little or no disproportionate influence on the competitive landscape. They are very vulnerable to competitive initiatives from smaller, more aggressive firms. Such changes can occur because of major technological shifts, through the development of proprietary technology, due to the deployment in new ways of existing technology as well as due to non-technology drivers. The competition tends to be among all firms in the market.

7.2.3.4   Knowledge Intensity

Knowledge intensity embodies more than mere technological sophistication and effective internal use of information technology. By knowledge, we mean information and data that has been captured, processed, and are accessible to the employees. Knowledge intensity is an indicator of how important all knowledge is winning in the marketplace. That knowledge may be knowledge of production processes, of customer preferences, or of supply chain dynamics.

In a marketplace where firms are very knowledge intensive, the technology base of the firms, as well as the technology that drives the underlying knowledge management processes, tends to be changing relatively rapidly. This allows the firms that are most knowledge intensive to gain, to hold, and to exploit competitive advantages, so long as they support that process.

In a marketplace where firms are only moderately knowledge intensive, the technology base of the firms, as well as the technology that drives the underlying knowledge management processes, tends to be relatively stable. While this allows the firms that are most knowledge intensive to gain, to hold, and to exploit competitive advantages, other firms can quickly catch up with, or even overtake, these firms.

In a marketplace where firms are not knowledge intensive, the technology base of the firms tends to be very stable. This provides no individual firm with any permanent competitive advantages, so any firm can quickly catch up, or even overtake any other firm.

7.2.3.5   Reliance on Alliances

The extent to which firms in your market rely on alliances, whether for research, manufacturing, distribution, or new initiatives, changes the nature and/or number of competitors. In fact, creating and exercising some control over complementary products, one frequent reason for such alliances is a well-recognized competitive strategy option.

Take, for example, a small software firm that is developing food service supply-chain software to compete with your firm's software. As a direct competitor of yours, the way in which it can and will compete will change rapidly in several dimensions in each of the following situations:

- The firm establishes a joint venture with a software firm which offers warehouse management software;
- The firm signs a development contract with one of the four largest food service firms, one which controls over 25% of the food service distribution market; or
- The firm agrees to partner with, say, Oracle and use only Oracle compatible modules in all of its new software products.

A market in which competing firms rely heavily on alliances is one where the actions of the firm with which you are competing are impacted by the terms of the alliance(s), the resource, and time demands of the alliance(s). Each firm moves, in a sense, with an impact including that of its allies.

A market in which competing firms rely only moderately on alliances is one where the actions of the firm with which you are competing are only indirectly impacted by the terms of the alliance(s), and the resource and time demands of the alliance(s). Each firm moves with its own impact slightly enhanced by that of its allies. Its own interests and a relatively independent strategy still largely drive its overall actions.

A market in which competing firms never, or rarely, rely on alliances is one where the actions of the firm with which you are competing are only occasionally, and usually, indirectly impacted by the terms of the alliance(s), and the resource and time demands of the alliance(s). Its own interests and an independent strategy still largely drive its overall actions.

7.2.4 Production and Supply Chain Issues

As firms seek to control and to redefine their own supply chains, as well as the way they produce their goods and services, they are changing the nature of the firms that you compete with. This is, in turn, divided into four separate areas:

- Level of technology;
- Relative number of suppliers;
- Degree of capital intensity; and
- Level of innovation.

7.2.4.1 Level of Technology

The level of technology that characterizes your marketplace should be viewed both in terms of product/service as well as in terms of the entire supply chain.

By production technology, we mean everything from the relative trade off between manual and automated production to the ability of one facility to produce multiple outputs with minimal changeover costs, including lost time. By supply chain technology, we mean everything from the ability to track the status of individual packages being handled by a carrier to the use of the Internet to distribute software by downloading.

In a marketplace characterized by high levels of production and supply chain technologies, we can expect to see competitors able to control and even reduce production and distribution costs on an ongoing basis, together with the rapid development and deployment of new products/services, as well as relatively low barriers to exit, but not to entry.

In a marketplace characterized by moderate levels of production and supply chain technologies, we can expect to see competitors able to control or eliminate increases in production and distribution costs on a short-term basis, as well as moderately rapid development and deployment of new products/services, together with relatively low barriers to exit as well as to entry.

In a marketplace characterized by low levels of production and supply chain technologies, we can expect to see competitors unable to control or eliminate increases in production and distribution costs, plus relatively low development and deployment of new products/services, as well as very low barriers to exit as well as to entry.

7.2.4.2 Relative Number of Suppliers

The relative number of suppliers to the firms in your marketplace is important since it is a rough approximation of the relative power of these suppliers. For example, if there are only three suppliers of a basic ingredient in an industrial detergent which is made by 25 different manufacturers, these three suppliers can have a significant impact on the market. So, if one supplier decides that this market niche is not profitable enough, it could raise its price or even leave the market. In either case, the firms that currently rely on that firm face a difficult competitive situation. If the price is increased, they may elect to absorb the price or to seek alternative sources. But what is the likelihood that either of the two other supplier firms have the capacity or inclination to supply that firm at the old price? Similarly, if the supplier ceases operations, do the firms it supplies have realistic options? They may not.

So, in a marketplace characterized by a relative small number of suppliers of key ingredients, components or inputs, relative to the number of firms in the market, the suppliers have, potentially, great power. In addition, in such markets, the reasons for the number of suppliers are often technological, such as that the technology used by the suppliers is proprietary, or financial, i.e., the production of the inputs may be very capital intensive.

In a marketplace characterized by a relative balance between the number of suppliers and the number of firms in the market, the suppliers have, potentially, much less power. In such markets, the reasons for the number of suppliers are often technological, i.e., the technology used by the suppliers is proprietary, or financial, i.e., the production of the inputs may be very capital intensive. In addition, firms competing in this niche tend to have developed technological abilities allowing them to move freely from one supplier to another.

In a marketplace characterized by a relatively large number of suppliers the suppliers have, potentially, very little power. Firms competing in this niche tend to have developed technological abilities allowing them to move freely from one supplier to another. Regardless of that, they are usually able to play one supplier against another to obtain better terms, including prices.

7.2.4.3 Degree of Capital Intensity

How capital intensive your competitors are also drives the kinds of data sources you should be focusing on.

Firms that are very capital intensive tend to have a significant reliance on existing technology, which is imbedded in their investments. In addition, since large capital investments often take time to plan and then implement, these firms will tend to have a longer range view of the market. Their responses to market changes, such as price cuts, will reflect not only current marginal pricing of their output, but also the firm's long-term vision of its place in the market, as well as the financial resources available to support the firm's ongoing operations.

Firms that are only moderately capital intensive tend to have a less significant reliance on existing technology. That does not mean they do not rely on technology. For example, that technology may partially be imbedded in strategic partners or suppliers. Since they are not making large capital investments, these firms will tend to have a shorter range view of the market than firms that are very capital intensive. Their responses to market changes, such as price cuts, will tend to reflect current marginal pricing of their output, as well as the financial resources available to support the firm's ongoing operations.

Firms that are not capital intensive tend to put little significant reliance on their own existing technology. That does not mean they do not rely on technology. For example, their technology may be imbedded in strategic partners or suppliers. Since they are not making large capital investments, these firms will tend to have a much shorter range view of the market than do firms that are moderately capital intensive. Their responses to market changes, such as price cuts, will tend to reflect current marginal pricing of their output, as well as the financial resources available to support the firm's ongoing operations.

7.2.4.4 Level of Innovation

Levels of innovation in an industry impact production of products/services. They also impact the development of new products/services, as well as changes in the way in which the supply chain operates.

Firms that have demonstrated low levels of innovation tend to view the immediate future of a market as an extension of the immediate past. Their time horizon tends to be rather short-term in nature. They are also more often reactive rather than pro-active, whether the issue is product improvement or workplace modification.

Firms that have demonstrated moderate levels of innovation tend to view the immediate future as reflecting the immediate past, but expect that many elements of it can and will change. Their time horizon tends to be medium-term in nature. They are also less reactive than they are pro-active, whether the issue is launching a new product or changing their own communications infrastructure.

Firms that have demonstrated high levels of innovation tend to view the immediate future of a market as only one step toward a longer term, more radically changed future. In many markets, their time horizon tends to be long-term in nature. They are significantly more pro-active than they are reactive, whether the issue is terminating a line of business, even a profitable one, or entering into a new strategic partnership.

7.2.5 Marketplace Dynamics

By marketplace dynamics, we mean the way in which the marketplace appears to you and your competitors. The dynamics of each marketplace you face may vary from product/service to product/service.

The way in which we look at marketplace dynamics is divided into three (3) separate elements:

- The number of customers;
- The geographic scope of the market; and
- The nature of competition.

7.2.5.1 Number of customers

The "number" of customers is a relative concept. Obviously, there are markets where the absolute number of customers is very small, for example less than 100. One example would be the market for civilian cargo jets for commercial cargo carriage. And there are also markets where the absolute number of actual and potential customers runs into the hundreds of millions or more, such as for cell phones. As we use it, the number of customers is measured relative to the number of firms.

In market niches where firms compete for a relatively small number of customers, these customers individually generally can exercise great power over the competitors, and thus may be the objects of separate CI projects. And the change in the relationship between any given competitor and its customer, or customers, can change the dynamics in the marketplace very quickly.

In market niches where firms compete for a relatively moderate number of customers, these customers individually generally exercise only moderate power over the competitors. But tracking them individually is not necessarily as critical as it is to spot the change in the relationship between any given competitor and its customer, or customers. Those changes will only moderately impact the dynamics in the marketplace quickly. In such niches, the way in which each firm, and each customer, faces the marketplace and acts in the short-term has a greater impact.

In market niches where firms compete for a small percentage of a mass of customers, these individual customers individually exercise no power over the competitors. Rather, in such niches, the way in which each firm faces the marketplace, acts in the short-term, and responds to the actions of the other firms has the greatest impact.

7.2.5.2 Geographic Scope of the Market

By geographic scope, we mean the area that your competitors serve, not the area that you serve. Geographic scope impacts performance issues such as logistics, economies of scale in the production of products/services, and ability to respond quickly to global versus local threats.

Those markets where firms compete on a global basis tend to have some larger firms with a global portfolio of products/services. To support such scale, these firms must typically develop and/or acquire significant internal support for global operations, plan for, and respond to long-term trends as well as near-terms ones.

Those markets where firms compete on a regional basis, on the other hand, may have some larger firms, as well as medium-sized and small firms. In addition, while the overall strategic goals of the very largest firms will impact the market, the goals and nature of the medium-sized firms as well will impact the present and future directions of the market.

Those markets where firms compete on a very local level often tend to have only medium-sized and small firms in the market. The ways in which these firms perceive the market and the ways in which they can and do respond to each other's activities has the greatest impact on the nature of competition in that market.

7.2.5.3 Nature of Competition

By nature of competition, we mean the balance between competing on price and competing on the basis of features. By competing on price, we mean a commodity-type market, while competing on the basis of features is a market where products/services tend to be "customized". In the past, customization meant low levels of production. However, as technology improves, markets, which previously were considered as commodity-type markets, now may be segmented into niches that can be considered as involving mass customization.

In markets where competition is based exclusively, or almost exclusively, on price, the product/service provided by any competitor is regarded as a satisfactory substitute for that of another, so the sole, or at least dominating, determining factor is price. Commodity markets have historically included fungible products, such as wheat. Today, they can include services, such as group health insurance benefits as well, where price is often the single most critical element in making a sale.

In markets where competition based on a combination of price and features, firms face mixed competition. That is, the product/service provided by any competitor is a satisfactory substitute for that of another, while the products/services vary in significant ways, ways that are very important to the customers. So the customers make decisions considering both cost and the benefit to them of features found in one product/service, against those in another. In the service sector, home security systems, with a mix of human and electronic elements, meet this criterion.

In markets where competition based largely or exclusively on features, firms are competing based on customization, whether or not the product is actually made in response to an individual customer order. The products/services vary in significant ways, ways of great importance to the customer. Thus the customers make decisions primarily based on the benefit to them of features found in one product/service, but not found in another. This can include products, such as a portrait painted for an anniversary, as well as services, such renting a limousine.

Scoring Your Way to CI Data Sources

To determine where best to look for the data you should be collecting and using, use the discussion above to analyze the market or markets and competitors you now face. Then "score" your responses. If you are in several markets, score each of them separately.

Use the following grid to figure your overall score. For each area, check the box next to the best description (the grayed out areas do not apply) Table 7.1.

Now, transfer your totals to the next chart Table 7.2.

If you have indicated one or more columns seven (7) times or more, those are the types of intelligence data resources you should focus on. If you have more than one column with seven (7) points or more, the one with a higher score is the more important for what you are facing now. If you need additional research sources after using the first set, go to the second set.

Use the following table to determine the channels of data collection that the experienced CI professionals have found are best mined for raw data for the market types you are now facing Table 7.3.

Table 7.1 Scoring sheet

Competitive Environment			Col. 1	Col. 2	Col. 3	Col. 4
	Company Regulation	High	☐			
		Medium	☐	☐		
		Low		☐	☐	☐
	Entry Barriers	High	☐	☐		☐
		Medium	☐	☐	☐	
		Low			☐	
	Exit Barriers	High	☐	☐		
		Medium	☐	☐	☐	
		Low			☐	
	TOTALS					

Products/ Services			Col. 1	Col. 2	Col. 3	Col. 4
	Substitutes	Few		☐		☐
		Moderate		☐	☐	
		Many			☐	
	Regulation	High	☐			☐
		Medium		☐		☐
		Low			☐	☐
	Lead Time	Short		☐		
		Moderate		☐	☐	☐
		Long	☐	☐		☐
	Life Cycle	Short			☐	☐
		Moderate			☐	☐
		Long	☐			☐
	TOTALS					

Companies			Col. 1	Col. 2	Col. 3	Col. 4
	Centrality of Management	Centralized	☐			
		Mixed	☐	☐		
		Decentralized		☐	☐	
	Cooperation	Highly competitive			☐	
		Moderately competitive		☐	☐	
		High degree of cooperation	☐	☐		
	Concentration	High	☐	☐		
		Medium	☐	☐	☐	☐
		Low		☐	☐	
	Knowledge Intensity	High				☐
		Medium				☐
		Low			☐	
	Alliances	Heavy reliance		☐		
		Moderate reliance	☐	☐		
		No reliance	☐			
	TOTALS					

(continued)

Table 7.1 (continued)

Production & Supply Chain			Col. 1	Col. 2	Col. 3	Col. 4
	Technology Level					
		High				☐
		Medium				☐
		Low			☐	
	Number of Suppliers					
		Few				☐
		Balanced number				☐
		Many		☐		
	Capital Intensity					
		Very intensive	☐			☐
		Moderately intensive	☐	☐		
		Not intensive		☐		
	Innovation					
		Low			☐	
		Moderate		☐		☐
		High	☐			☐
TOTALS						

Marketplace			Col. 1	Col. 2	Col. 3	Col. 4
	Customers					
		Few		☐		
		Moderate number		☐	☐	
		Many			☐	
	Geographic Scope					
		Global	☐			
		Regional	☐	☐		
		Local		☐		
	Type of Competition					
		Price			☐	
		Price & features			☐	☐
		Features			☐	
TOTALS						

Table 7.2 Table of totals

Total number of hits
Col. 1
Col. 2
Col. 3
Col. 4

Table 7.3 Source summary

Typical sources of raw data	Col. 1	Col. 2	Col. 3	Col. 4
Secondary				
Trade journals	◆	◆	◆	◆
Trade associations and chambers of commerce	◆	◆		
Government reports	◆	◆	◆	◆
Government records and files	◆	◆		
Security analyst reports	◆	◆	◆	
Academic case studies	◆			
Research centers	◆	◆		◆
Business information services		◆	◆	
Advertisements		◆	◆	◆
Want ads		◆	◆	◆
Local newspapers and magazines	◆	◆	◆	
Competitive reports	◆	◆	◆	
Competitor news releases	◆	◆	◆	◆
Competitor home page	◆	◆	◆	◆
Directories and reference aids	◆			◆
Technical publications			◆	◆
Catalogs				◆
General news publications	◆	◆	◆	
Books	◆	◆		
Internet aggregators and portals		◆	◆	◆
Primary				
Your own employees	◆	◆	◆	◆
Industry experts	◆	◆	◆	◆
Sales representatives		◆	◆	
Customers		◆	◆	
Security analysts	◆	◆	◆	
Competitors (contacted directly)	◆	◆	◆	
Suppliers	◆	◆	◆	◆
Product and service purchases			◆	◆
Focus groups			◆	
Questionnaires and surveys		◆		
Trade shows		◆	◆	◆
Industry conferences	◆	◆		◆
Technical and professional meetings				◆
Facility tours		◆	◆	◆
Retailers and distributors	◆	◆	◆	
Banks	◆	◆		
Advertising agencies	◆	◆	◆	
TV, radio programs, and interviews	◆	◆	◆	◆
Speeches	◆	◆	◆	◆
Internet chat groups		◆	◆	

7.3 Starting with the Sources of Raw Data

The numerous sources listed in this section of the chapter are just starting points. While potential sources for raw data are almost endless, for ease of management, we divided the most likely potential sources of raw data into three basic categories:

- Government & Non-profits
- Private sector
- Media

Grouping these data sources allows you to keep track of them. More importantly, data sources within each category have important characteristics in common. Knowing these characteristics can be important when you begin your analysis because every data source is not created equal.

As you begin to focus on where the raw data you need might be located and accessed, remember that where you get data is not necessarily the ultimate source of that data. This means that you may be able to access the data you want without having to contact the data's original source. But, it also means that you must make sure you know whether a source of data actually produced the data, just transmitted it, or modified it in some way.

Confusing the data producer of data with the data provider has important, sometimes damaging, consequences. For example, cross-checking past estimates made by a trade association that prints market predictions with the industry's actual performance might show that this publication is often incorrect in its predictions. That may be because it is merely reporting the views of those in the industry. In such cases, treat it with care.

7.3.1 Government and Non-Profit Sources

As a group, Government and Non-profit sources generally provide only indirect assistance. That is because the vast bulk of the data that they access and release is highly aggregated, as with business census reports, or consists of data already collected by another provider, such as information taken from a commercial directory. Some, however, can be very specific, firm level or even facility level, such as filings made with a federal, state, or local government by a target. Data in those channels tends to be very company- or subject-specific rather than aggregated. It tends to be relatively easily and inexpensively accessed whether online or through freedom of information/open records laws, although it may take a relatively long time to do so.

Examples of Government & Non-profit Data Sources

US Government
    Regulatory Agencies
    Trade Promotion Offices
    Congressional Hearings
    Winning Competitive Bids files
    Court Cases, Dockets, and Records
    Patents & Trademarks
    Agency & Contractor Studies: Regular and One-time
Foreign Governments
    Regulatory Agencies
    Trade Promotion Offices
    Patents
    Commercial Attachés
State Governments
    Regulatory Agencies
    Court Cases, Dockets, and Records
    Winning Competitive Bids files
    Environmental Permits and Other Regulatory Filings
    Trademarks
Local Governments
    Zoning & Building Permits and Other Filings
    Court Cases and Records
    Winning Competitive Bids files
    Industrial Development Authorities
    Boards of Taxation and Assessment
Chambers of Commerce
    Domestic
    Foreign Chambers in the United States
    US Chambers of Commerce Abroad
Consumer and Other Advocacy Groups
    Non-Government Organizations (NGOs)
    Product Tests & Comparisons
    Regular Publications
    One-time Studies & Position Papers
Trade Associations
    Regular Publications
    Membership Directories
    Special Studies and Reports
    Meetings & Reprints of Speeches
    Statistical Abstracts
Academics and Academic Resources

Faculty
Regular Publications, Special & One-time Studies
Industry Research Centers & Specialized Libraries
Teaching Materials & Case Studies

The Consumer and Other Advocacy Groups and the Trade Associations all collect and provide data for a reason—to advance what they each see as their own best interests or the best interests of those that they represent. To put it bluntly, they all have an "ax to grind." In doing this, they may well spend significant time and funds to collect data, publish reports, bring lawsuits, or test products/services, all of which may be sources of raw data for you. However, some of them may limit their data to members.

From the CI research perspective, Academics often seek funding support for research which they are interested in, or conduct consulting assignments. From these efforts, the professors and their researchers may provide such useful input as publications, special detailed studies, and access to research centers for collecting important historical data.

7.3.2 Private Sector Sources

This includes people and organizations whose business directly involves producing or selling the kinds of data you may be seeking. For some, providing the data is their business. Others may come across data you need as a part of their own business.

Examples of Common Private Sector Data Sources

Your Primary Competitor's Employees
  Sales
  Market Research
  Planning
  Engineering
  Purchasing
  Former Employees of the Target Company
Your Primary Competitor
  Internet Home Page
  Catalogs & Price Lists
  In-house Publications
  Press Releases and Speeches
  Presentations to analysts
  Advertisements and Promotional Materials
  Products
  Annual Reports
  Regulatory Filings

Customers & Suppliers
  Retailers, Distributors, and Agents
  Advertising and Marketing Agencies
Other Competitors
  Business Information Services
  Dun & Bradstreet
  Standard & Poor's
  Proprietary Research Firms
Experts
  Consultants
  Expert Witnesses
  Security (Stock) Analysts

In dealing with these sources, you must avoid confusing the package with its contents. For example, let us say that you have purchased a proprietary research report on a targeted competitor and are now reviewing the data it provides you on the firm's size, employees, sales, and so on. To verify the data, you may compare it with data you have obtained from another business information source, such as a business profile report. Suppose the facts appear identical. This may appear to provide confirmation of the first set of data. However, that is not necessarily correct. You see, the report's author could have purchased the data on your competitor, so if it looks the same as data from another business source, that may be because it is the same. That does not mean it is correct. This is just a false confirmation.

Experts include everyone from consultants to expert witnesses, and from clinical laboratories to security analysts. Their work reflects a common goal: to advance the individual's career, whether it is by obtaining assignments, helping an employer sell stock, or some other means. But each is providing data for a particular audience, which can color not only how they say things but what they say and do not say.

7.3.3 The Media

All these varied sources collect, generate, and process data for a specific audience. To fully understand both the data you may find on your competitor and how to analyze it, you must first understand from whom the media collects that data, how, and why it releases it.

Examples of Common Media Data Sources

Business Newspapers and Magazines
  Advertisements & Want Ads
  Articles
  Reporters
Wire Services
  Articles

Reporters
Specialized Directories
Local and National Newspapers
Advertisements & Want Ads
  Articles
  Reporters
  Obituaries
Technical Journals
  Articles
  Authors
Trade Papers and Journals; Financial Periodicals
  Advertisements & Want Ads
  Articles
  Reporters
  Marketing Studies & Media Kits
  Special Issues
  Related Publications
Security Analysts' Reports
  Company Reports
  Industry Profiles

The media, in the broadest sense, can be one of the most productive resources for obtaining raw data. But, always keep in mind that many publications exist to serve a particular industry or market. Thus they are positioned to help you locate some important data and develop leads for additional data, but may their own significant blind spots as well.

As one US government intelligence manual dryly notes,

> Secondary sources such as government press offices, commercial news organizations, NGO spokespersons, and other information providers can intentionally or unintentionally add, delete, modify, or otherwise filter the information they make available to the general public. These sources may also convey one message in English for US or international consumption and a different non-English message for local or regional consumption....
> All media are controlled. The issue for analysts is what factors and elements (elites, institutions, individuals) exercise control, how much relative power or weight does each factor or element possess, and which factors or elements are of interest to analysts and their customers. (US Department of the Army (2006), pp 2–10, K-1)

7.4 Planning for Research You Will do Later

You will probably face the possibility that you will have to conduct research against a specific target in the future, but you do not know when that will happen. In our experience, it is useful for you to develop a summary of the potential target, and, most importantly, where you can locate data on it if and when you need it. Consider adapting the following form to memorialize that sort of advance work

Target:
Ultimate Parent Company:

Identification Information
Address/Telephone/Product(s):
Does it make filings With the US SEC or other
regulatory agency?
Target Data Sources
Website(s):
SEC Filings and Related Releases: Yes/No
Comments/Suggestions:
Newsletter(s): Yes/No
Name:
Frequency:
Contents:
Distribution:
Comments/Suggestions:
Table of Organization:
Commercially Available:
Advertising Agencies
Public Relations
Internal
Contact:
Coverage:
Comments/Suggestions:
External Firm
Name:
Address:
Contact:
Coverage:
Comments/Suggestions:
800/Consumer Information Number(s):
Business School Cases:
Other:
Overall remarks:

This form actually captures some preliminary research efforts for the next time. In doing this, you spend a limited amount of time determining where information on your target will probably be when you eventually need it. You do not collect it now. However, if and when you need to start collecting that data, you will already have an idea of where it might be, and just as important, where it is not likely to be.

References

McGonagle JJ, Vella CM (2002) Bottom line competitive intelligence. Quorum Books, Westport, CT

US Department of the Army (2006) Open source intelligence. http://www.fas.org/irp/doddir/army/fmi2-22-9.pdf. Accessed 11 Oct 2011

Chapter 8
Research: Specific Cases

8.1 Secondary Versus Primary Research

You will find that you are limited to using a combination of secondary research plus whatever primary research you can do yourself. The reasons for this vary widely. For example, a project may have severe financial constraints, thus limiting your ability to arrange for time-consuming interviews. For whatever reasons, primary research is often not used as much as it should be. Do not make the mistake of ignoring it.

Primary research, particularly "human intelligence", is often misunderstood or even held in low esteem. Some erroneously see it as inferior to other forms of data gathering, particularly that which is based on "hard" (text) sources, such as scientific journals. It is neither inferior nor superior. It is just different in several key aspects:

- Human intelligence results tend to be approximate, while hard copy sources are, or at least appear to be, precise;
- Human intelligence allows only indirect access to data, while print sources may provide direct access; and
- Human intelligence can often only supplement primary data already gathered from other sources.

But, human intelligence can have significant strengths, such as:

- It tends to deliver more present and future oriented data, while print sources tend to be past-oriented;
- It is unfiltered, while print sources may be highly filtered. That filtering is due to content control, by whatever name, e.g., editing, spinning, or disinformation;
- Print-sourced intelligence is more likely to be subject to timing limitations e.g., publication lead times, distribution embargoes; and
- Human intelligence sources provide an easier opportunity to develop additional data or identify additional sources of data than do many hard copy sources.

J. J. McGonagle and C. M. Vella, *Proactive Intelligence*,
DOI: 10.1007/978-1-4471-2742-0_8, © Springer-Verlag London 2012

8.2 Secondary: Internet Tips and Rules

The Internet should be viewed as a step toward getting data, *but not* as the ultimate source of all of your data. By that, we mean that you can get information faster from secondary sources on the Internet than using other options—but only if that information is already there.

One of the real values of the Internet is the ability to use business and social networking groups to identify and even contact potential interviewees. For example, if you are trying to identify someone who had previously been an intern at a particular firm, going to a social networking page may enable you to find her/him. Similarly, using business networking pages not only helps you identify individuals at a target for possible contact, but, more importantly individuals who previously worked with the target. A former employee certainly does not always have current information, but he/she almost always have information that is more valuable and current than that which is found in dated print sources.

The Internet is a growing, morphing place. By the time you read this book, it will have grown substantially and it will continue to change rapidly. Rather than try to give specific search tips that will become obsolete as soon as they are written, we want to give you a set of general rules to make sure that you understand what is happening:

Rule 1. Not everything you need is on the Internet for free.

Rule 2. Not everything that has been published on the Internet for free.

Rule 3. Treat the Internet with caution. What *is* there is what people, firms, and other institutions want you to see. That does not always mean it is accurate. For example, if you look at a profile on Facebook or LinkedIn, you are reading what the person that created his/her profile wants you to read. If it is important to you that the information be correct, do independent checking.

Rule 4. Not everything on the web can be found using any, or even all, available search engines. At present, Google dominates the search engine business, but Microsoft's Bing search engine has some advantages, particularly when you can phrase your need in the form of a question which is likely to match the heading on a document or table. In many cases there is hidden content, not identifiable through search engines:

- For example, while a local zoning board may make access to copies of zoning applications available to Internet users, these documents will not be indexed by a search engine. This is because you have to go to the municipality's site, and then search for the data behind its protective wall. This puts the pressure on you to think about where things might be and look there for them.

- Another form of hidden content can be very revealing in developing your CI:

Recently we were looking at the website of the online retailer [name deleted]. They had announced that they were about to introduce a new "home & garden" section to the site. We found that this section was actually available with full details as a hidden link and was obviously being tested by [the firm] prior to launch (Weiss and England 2000).

Rule 5. Just because something is on the Internet today does not mean it will still be there tomorrow. So, if it is important to have repeat access to something on a webpage, or you may have to go back to it for further reference, download it, save it, grab it, copy it, for tomorrow it may not be there. This issue is sufficiently sensitive that lawyers are now looking for ways to prove that a particular Internet page looked like this and said certain things on a particular day when preparing for trial that might be years later.

Rule 6. Some things that are no longer on the Internet are still there. Here we are speaking of the value of learning about caching and archiving. As Google describes its own caching:

> Google takes a snapshot of each page it examines and caches (stores) that version as a backup. The cached version is what Google uses to judge if a page is a good match for your query. Practically every search result includes a *Cached* link. Clicking on that link takes you to the Google cached version of that web page, instead of the current version of the page (Google guide 2011)

Archiving, on the other hand, is conducted on sites such as Archive.org:

> The **Internet Archive Wayback Machine** is a service that allows people to visit archived versions of Websites. Visitors to the Wayback Machine can type in a URL, select a date range, and then begin surfing on an archived version of the Web. Imagine surfing circa 1999 and looking at all the Y2K hype, or revisiting an older version of your favorite Website. [Archieve.org (2011)]

Rule 7. Be careful out there. Even if you remember to use tools such as "private browsing" *every time* you use your browser, you will leave traces of your visits to websites. Also, make sure your virus protection is up-to-date and always on. Finally, keep in mind that your own employer may be tracking your Internet usage, so let your IT people know that you are doing your own CI.

Rule 8. On the flip side, while it is not yet commonly used, there is no reason why your competitor's website cannot divert some user traffic, say from your firm, away from sensitive areas based on a user's ISP address:

> Indeed, one trick...involves domain-name identification, in which Websites are configured to detect browsing competitors by their domain names, and divert these visitors to special pages. Cisco Systems redirected such visitors to a page containing holiday-party scenes, then shunted them to a page about employment opportunities at Cisco. (Kahaner 2000)

Rule 9. When faced with an overwhelming numbers of "hits", think of different ways to narrow the search other than adding more terms. For example, are there likely to be business documents that have been circulated outside the target firm? If so, consider limiting your search to PDF files. Has the firm likely made presentations to potential customers, suppliers, etc.? If that's a possibility, try limiting your search to PowerPoint files.

Rule 10. Just because you find the same information on several sites no longer means that the first bit information is now confirmed. In the Internet age, where copyright and ownership rules are increasingly flouted, it may well mean that several sites have "adapted" the same content or merely reposted it. In other words, quantity is not the same as quality.

Rule 11. Every once in a while, when you find something particularly interesting on a competitor, take a few seconds and see whether you could find the same kind of data or resource on your own firm. You can make yourself extremely valuable by pointing out such vulnerabilities to your peers in CI or in corporate security.

Rule 12. When exploring blogs, remember that you have no idea who is actually making a posting, or why they are doing that. Even when a blogger posts what he/she claims to be a photo of a forthcoming product, proceed with extreme caution. There are cases where major corporations have pursued bloggers for releasing, or trying to release, information on forthcoming products. What the blogs can be more useful for is to monitor consumer opinion.

Rule 13. When dealing with blogs, particularly blogs that deal with a particular firm, be careful about entering the discussions. Increasingly, firms monitor the blogs that focus on them, so that the appearance of unusual questions or provocative statements may alert them, if not to your firm's interest in them, at least to the fact that there is unusual interest in them out there.

Rule 14. Do not rely on the blogs for hard data; use them to get to hard data. Despite what you hear, blogs are not just opinion pieces but for example, in many cases, opinion pieces based upon research studies. What you are interested in are the research studies, the hard data. You can find these using the regular search engines, such as Google. When you look at the Google page, on the left side, there is a listing for Blogs. Go to that and enter your search term. Let us use the example "natural cereals". We immediately got the return of reports that have been completed, when we did this, within the week. Of course there were blog remarks, but we were not interested in those answers. What we did was to follow the links which led us to the private company which did the report. So, instead of just getting a summary of the report and the press alert, we finally tracked down the report itself with the extensive data the blogs only referred to.

Rule 15. The Internet can be an incredibly useful place for locating potential interviewees. Not only can you find useful leads by using a site such as LinkedIn to find former employees of a target, but searches of Facebook, and similar social sites, can be useful in providing the odd details. For example, in one case, we were able to confirm a target firm was building up a particular specialized team. We did that from notes in a Facebook page of a newlywed who was bragging that he was returning to a much more important job with the team. He proudly noted that he was going to be part of an effort to double its size. Also, there are sites that purport to provide "experts". More often than not you will find them associated with sites that serve litigators. Before using them, realize that experts make their living selling their information, not giving it away. But if you need to turn to such a person, these sites can help to provide leads.

Rule 16. Not everyone keeps everything up-to-date every day. Just because someone's profile seems to indicate that they are looking for job, or you find that they have posted a resume, does not mean that they have not already gotten another job. They may just have forgotten to take the resume down. Conversely, individuals who have recently lost their position very often will not make a corresponding change in professional sites such as LinkedIn in order to appear more attractive to recruiters or potential employers.

Rule 17. Just because a government agency, whether federal, state or local, has a site where you can request information under the US Freedom Of Information Act or similar state or local law, does not mean the process is now faster than before the agency went online. In many cases, these sites merely speed up the initial communication of your request, allowing it to be made electronically rather than by mail. However, the back-office work is still just as slow, in some cases, almost excruciatingly so. On the other hand, there are some government agencies, for example a few state insurance departments, that post much information which can be obtained directly online, which is still provided only on a manual search basis in other states, such as policy form filings.

Rule 18. Do not be seduced by the fact that a tremendous amount of secondary data can be located from the Internet now. Secondary data is just that— secondary. Too often people not experienced in collecting intelligence fall victim to the illusion that they have everything that they need, because everything that is collected on the Internet seems to answer their questions. Do not fall into that trap. Use the Internet as a step toward getting primary data, that is, to complete all of your background research so that you have mastered a particular subject before you seek to identify interviewees and then approach them.

Rule 19. Just because you cannot find it on the Internet does not mean it is not there. It may be useful for you to seek help from someone particularly skilled in a particular area when trying to conduct a search. Just as you would want someone who was an expert on Westlaw Next, a primary source research service for lawyers, to run a search on that service, never hesitate to ask for help in terms of running or designing a search on the Internet. Google alone has a constantly increasing variety of places and ways to search. No one, regardless of their expertise, has mastered all of them.

Rule 20. Providing a report of what you found on the Internet is not intelligence; it is not even analysis (Gonsalves 2008).

Rule 21. Not everything is on the Internet.

8.3 Primary: Working Trade Shows and Industry Conferences

Why should you bother working a trade show or industry conference? Listen to what just one individual can do:

> I collected on three major competitors and two minor ones. I conducted about 10 interviews and attended a cocktail party hosted by my company's key competitor. My company wanted to learn tactical details on each of the competitors, including the sizes of the various paneling systems. Additionally, I collected information they never dreamt would be available from their key competitor: which new products the distributors thought were hot or not and why, and how they planned to sell them (Calof 2008).

Collecting CI at a trade show or industry conference has a few twists in terms of collection as well as in terms of analysis.

In terms of collection, working a trade show or industry conference it has several major differences from other activities in which you might engage:

- First, there is heavy use of elicitation techniques to develop data from individuals at the event. That does not mean that you should ignore other sources of data. Listening to how a speaker answers questions, either at the podium or in a crowd can be invaluable. But a trade show or industry conference is also an event that should generate paper, CDs, samples, presentations, advertising materials, attendee lists, etc. Do not ignore the secondary collection. You have only one chance to get it.
- Second, you need substantial preparation to be able to work the event, since it is, by definition, limited in time. That often forces you to have to select a few from among many potentially useful targets, unless you can recruit others to help you.
- That takes you to the third difference, that is, the people working the trade show have to have or have to be given a good background on the technology/subject matter of the meeting. That lets you and them be able to "swim" in that environment as soon as you all arrive.
- The last is the need, in some cases, to use outsiders. That is because, at most trade shows and industry conferences, your competitors may (should) know who you are and that you might be conducting CI. Thus, you may want to bring in third parties to help you with your data gathering.

In terms of analysis, the different issue is associated with the need for prompt debriefings. That means getting everyone who worked the event, whether or not they were engaged in collecting CI, to exchange all they learned as quickly as possible, preferably in a group with you and any other CI collectors. Experience shows that this should happen before, not after, getting back to work. At that time, you should not only know what you have found, but should also determine if there was a dog that did not bark. That is, what did you not hear that you expected to hear about?

8.4 Primary: Telephone Interviewing

Telephone interviews can be a powerful way to supplement your secondary research. However, really effective telephone interviewing can be difficult for a novice to master. Before undertaking it, first think about these three points:

- Preparation
- Interview strategies
- Techniques to keep an interview moving.

8.4.1 Preparation

How you will identify yourself? Never misrepresent who you are. For example, do not say "I'm a student." There are several problems you face in misrepresentation:

- Foremost is the ethical issue. This is just unethical. Period.
- This kind of unethical behavior is actually counterproductive. Over time, engaging in "just a little" unethical behavior feeds into a downward spiral in your own behavior.
- Misrepresentation can result in data contamination. That occurs when the interviewee provides the type of information he/she believes this specific interviewer needs or wants. Thus, someone pretending to be a student may get student-level, assimilated data.

You are obligated to say who you work for, but is there an obligation to make excessive disclosures? No, so long as what you say is honest.

Be clear in your own mind what you want to accomplish in the interview. Specifically, what types of data do you expect the interviewee to contribute? Decide whether you want to ask focused questions or broad-ranging ones. The former provides more specific, statistically usable data, but most people feel more comfortable dealing with broad, open-ended questions—especially if their answers draw on personal and professional experience. In either case, keep your questions as short as possible.

Consider such mechanical issues as note-taking, the interviewee's use of caller ID and getting calls returned:

- Will you use a script? You really should not as you will sound artificial in what should be a casual conversation.
- How will you make notes? Will the interviewee let you record the conversation? Remember, recording a telephone conversation without the consent of the other party is almost always illegal. Also, asking to record it may make your interviewee feel uncomfortable, or even decline to talk with you. So get used to taking good notes.

- Do you want your interviewee to sense or know that you are taking notes? Can you wait until after an interview and write-up everything that is important?
- If caller id may be in use, the number displayed to your target may impact whether or not your call is even taken.

8.4.2 Interview Strategies

Having determined what you want to get from an interview (or interviews), next establish your own strategies for:

- Identifying, and then dealing with time constraints
- Approaching your targets, and
- Dealing with sensitive targets.

8.4.2.1 Time Constraints

Set your own deadline for conducting interviews and then stick to it. Unless you do that, you can quickly get enmeshed in an endless round of interviews that provide decreasing amounts of useful data while using up your valuable time.

Make sure that you clearly understand the following:

- How much time do you have available? The answer has at least two different aspects. The first is how much time do *you* have from the beginning of the interviews until they must be completed? Second, how much time is available is available *to do* the interviews at any one time?
- Given the length of time available, just how available is that time? For example, have you allowed sufficient time for preparation (and practice if needed)? How much time did you allow for the calls themselves? Will you be available when interviewees may call back? Did you allow time for follow-up calls to complete outstanding interviews? How much time is allotted for completing reports on calls and then analyzing the results of each completed interview?

You should limit the number of interview calls per day to no more than 3:

- It is important to pace yourself. It takes time to prepare for an interview, to complete an interview, and prepare your notes. Experience shows that, for every hour on the telephone, you will need from a half-hour to an hour to prepare for an interview, and then to identify and make contact with a potential interviewee, and up to another full hour to write-up that interview.
- By limiting the number of outgoing calls, you can more easily set aside a specific callback time, if needed. You also minimize the likelihood that you will miss a returned call. If you have a call waiting signal, you might consider disabling it so you do not interrupt an interview in process.

8.4.2.2 Planning an Approach

How you will approach your interviews? First, determine what your opening will be. That is, why are you calling and what are you looking for? In general, keep your opening short and practice it, if necessary.

Consider how you will talk about what you are interested in and conduct the interview. For example, are there technical terms, key words or buzzwords that the interviewees are likely to use? If so, you should identify them and learn what they mean first. This does not, however, mean that you should use them in the interview process. In fact, there are several situations when it is not appropriate to use these terms:

- When you are not fully conversant with them, because you might make a mistake in using them, or in translating them in your notes.
- When using such terms gives your interviewee the belief that he/she is dealing with an expert. That could limit the background or context that he/she gives you, which may be exactly what you are seeking in the first place.

However, you may want to use such terms when you deal with a gatekeeper. That is, if you do not know who your interview target is, but you do know in what area he/she works, having command of these terms makes it easier for a gatekeeper to connect you to the right person.

8.4.2.3 Sensitive Interviews

There will be situations when you have to approach an interviewee who is "sensitive". By sensitive, we mean someone who:

- Is difficult to reach,
- Has a limited amount of time to talk with you,
- Has a great deal of expertise, so is likely to refuse to talk to you for very long, and/or
- May be uncomfortable talking with you (or anyone in your firm).

In these cases, you should do as much work as possible before starting the interview to maximize the value of the limited time you may have. For example, if you are dealing with a high-profile expert, exhaust all public sources first. Make sure you know what the person has already said or written so you can seek something new or different and not waste his/her time (and yours).

Also, be prepared for a hand-off, that is the interviewee's desire to have you talk to someone else, or at least to get rid of you. Because of that, always get a name and a direct dial telephone number, if possible, as soon as you see this happening. Try then to ask your interviewee if there is anything he/she can add to what that source will probably say. In other words, exploit who you have on the line while you have him/her.

8.4.2.4 Keeping the Interview Moving

There are a wide variety of techniques available to keep an interview moving. All are based on understanding the interviewee, mastering the subject in question, and your patience. What follows are several of those that have most often been overlooked.

When you start, do not say you are "seeking information" or "doing a survey". This can put off your interviewee, and quickly result in him/her transferring you to someone else or just terminating the call. Rather, approach the situation more personally, pulling the interviewee quickly into a discussion with you. Consider opening approaches such as "I'd like to get your opinion on this." or "Do you have any thoughts (experiences) on that you can share?"

Prepare yourself, perhaps by a rehearsal, to anticipate problems (or objections) and to have ready answers for them. For example, always be prepared to deal with the following common objections:

- "I'm too busy now."
- "I really don't know much about that."
- "I can't talk about this."
- "Why are you calling?"
- "You'll have to talk to [name] about that (but she's not in)."
- "What are you going to do with this information"
- "We do not give out information (like that)."

Approach the interview in an orderly manner. Before you start, remember the memory trick for writing newspaper stories: "Who, What, Why, Where, When, and How." For you, this means that your interview should ideally cover all of these. So should your post-interview notes.

When talking, it is usually better to start with the easy subjects first and move to hard ones, or to move from the general to the specific. This way, you and the interviewee move naturally through what should always be a discussion, never an interrogation.

If there are subjects that the interviewee may find sensitive or even objectionable, keep them for the end of the interview. That way you will have extracted as much as you can for the interview before your subject decides to terminate the discussion.

8.4.2.5 Adopt the Right Attitude

When you approach the interview, convey the attitude that you are seeking a "chance" to get some information. Do not try to get it all at once from only one subject. Often that effort will result in you getting none at all.

When covering the points you wish to focus on, speak less rather than more. Silence literally is golden. Experience demonstrates that if an answer you receive is not enough, a careful pause, that is sheer silence, can often result in the subject resuming the conversation and trying to elaborate to avoid the awkward pause.

When conducting an interview, make sure to listen for what your interviewee knows (and what he or she does not know). Specifically, you should be ready to do the following:

- Modify your questions to fit your interviewee's demonstrated level of knowledge.
- Broaden your questions by using his/her answers as a springboard.
- Use challenging statements instead of questions, for example, "Well, there are reports that production is down over 15%." Such a statement may elicit a quick, and revealing, rebuttal.
- Work at inducing clarification and cooperation. For example, repeat what you just heard instead of asking a new question. Also avoid being the source of a potentially controversial question or statement. That means you might preface a sensitive topic with, "Some in the industry feel that", so your interviewee knows that this is not your opinion and does not react negatively to you.

8.5 Primary: Dealing with Former Employees of Competitors

A very useful source of intelligence on a competitor can come from interviews with individuals who formerly worked there. And the easiest of those for you to access will be those who now work for your firm.

Does that mean you can just go up to them and start interrogating them? Of course not. There are several rules you should adhere to when dealing with them.

Realize that the more time that has passed since an individual left a competitor, the older their data will be. That is true even though most people, when asked about former employers will answer in the present tense.

Before asking an employee about his/her former employer, be careful. Get permission from his/her supervisor to do so. But, make sure to alert the supervisor that you will not be discussing what the employee says with the supervisor.

Approach the employee carefully. Identify yourself and tell them why you want to talk with him/her about the former employer. Before continuing, tell them this (and always respect it): "I do not want to have you tell me anything that is a trade secret or confidential. I do not want you to answer any question that makes you uncomfortable. If you signed any kind of a non-disclosure agreement with that employer, I do not want you to violate. Also, I will not tell your supervisor anything except that you were cooperative."

When talking with the employee, do not be pushy, do not challenge, and do not pursue topics that clearly make the individual uncomfortable. If he/she does not want to talk with you, offer your thanks and move on. If the individual wishes to cooperate, at the end, you may ask if he/she knows another individual who formerly worked at the target, but who is no longer there, and who might know about the subject that interests you. If you do get a name, approach that individual the same way, even one who does not work for your firm.

When referring to what the employee says in any context, never, ever, use his/her name. Describe him/her only in general terms such as "a former sales employee". If you can avoid it, do not mention whether or not he/she works for your firm. The reason for this caution is that you never want to put an individual in a position where they fear that a failure to cooperate could cost them a job. You are better off avoiding such interviews if there is any chance that might happen, in spite of your best efforts.

References

Archive.org (2011) The wayback machine. archive.org/about/faqs.php#The_Wayback_Machine. Accessed 8 Aug 2011

Calof J (2008) Conference and trade show intelligence. In: Presented at the 2008 international annual conference and exhibition of the society of competitive intelligence professionals

Gonsalves A (2008) Despite the internet, google generation lacks analytical skills, Information-week.   http://www.informationweek.com/news/205901358?queryText=Analytical+Skills+GONSALVES+INTERNET+GOOGLE+S. Accessed 11 Oct 2011

Google (2011) Google guide: making searching even easier. googleguide.com/cached_pages.html. Accessed 8 Aug 2011

Kahaner L (2000) Keeping an 'I' on the competition, Informationweek. http://informationweek.com/805/main.htm. Accessed 25 Sept 2000

Weiss A Steve E (2000) Internet intelligence—analysing web-sites for competitive intelligence. http://www.freepint.co.uk/issues/220600.htm#tips. Accessed 16 Oct 2000

Chapter 9
A Deeper Look at Data Sources

The data that goes into supporting your Proactive CI analysis may be found virtually anywhere. However, CI's 25+ years experience is that there are only so many places where potentially useful data can be accessed legally and ethically. And, the collective experience of CI analysts is that there are places where one can usually expect to find useful data and places where it is extremely unlikely to find that data.

Based on that experience, we have outlined the most common, typical sources of raw CI data. For each, we have, very briefly, indicated what kinds of data you can expect to find there.

Data sources for CI research can be divided into two broad, separate groups: secondary and primary:

- By Secondary, we mean sources that are primarily prepared for the public to read or to see.
- By Primary, we mean sources that have data, but which are not prepared for publication.

On the whole, the Secondary sources are in print and Primary involves inter-action with people or non-print items. But the line between them is not always clear. While a newspaper article may be considered secondary, how would you classify a telephone call to the reporter who did the story? Does it depend if he/she relates additional, unpublished but public data collected from the Target? What if he/she provides you with names of potential interviewees at the Target?

For each broad type of data source, we will briefly tell you the following:

- Audience: To what audience is the document, the publication or the contact devoted or aimed?
- Purpose: Why was this source created, why was this data provided or why was it originally collected?
- Sources of underlying data: Where did the data provided actually come from. For example, did it come directly from a Target and if so, from someone at the Target who would be expected to have that data?

J. J. McGonagle and C. M. Vella, *Proactive Intelligence*,
DOI: 10.1007/978-1-4471-2742-0_9, © Springer-Verlag London 2012

- Level of Detail and Scope: How much depth or breadth can you typically expect to find in data at this source? Would it be considered as superficial to individuals in your position, or as having a great deal of depth? Would it be considered as very narrow and focused, for example on production details, or would it be written for a general audience?
- Age: How old is the data provided here likely to be? Is it very current, that is less than 1 week old, current, that is less than 1 month old, dated, that is up to 12 months or which is over 12 months old?
- Accuracy and Completeness: How likely it is that data obtained from or through this type of source is accurate as well as complete?
- Knowing these will enable you to more quickly decide whether or not to go after data from a particular source, or whether it is likely to be a waste of your time.

9.1 Secondary Sources of Raw Data for CI

9.1.1 Trade Journals

Audience: Employees and executives of firms in the industry; suppliers, and customers, to a lesser degree.

Purpose: Keep readers up to date with news in the industry and factors impacting that industry niche. Often includes interviews on future developments, trends, and strategies.

Sources of underlying data: Companies in the industry, trade associations, government agencies, wire services.

Level of Detail and Scope: Usually highly focused on industry, assuming its readers understand who is who and what are the major issues, technologies, etc.

Age: Materials are usually fairly current. As a general rule, online publications have materials 1–3 days old; weekly publications are 1–2 weeks old, monthly publications, 1–2 months old, and annuals are often 6 months old before they are published.

Accuracy and Completeness: Tend to be fairly accurate and complete. Only problems tend to be in publications that have a high turnover in editorial personnel, which produces a degree of loss of institutional memory.

9.1.2 Trade Associations and Chambers of Commerce

Audience: Publications tend to aim at senior-level management and other key decision-makers in a particular industry, or local businesses.

Purpose: To communicate changes affecting in the overall industry, or local economic conditions, and to marshal support from members for legislation, research and development, etc.

Sources of underlying data: Members, particularly their public relations staff and consultants.

Level of Detail and Scope: On matters impacting the industry, deals with issues at more common, i.e., strategic, level. When dealing with individual members, can sometimes provide a detailed profile of people and processes.

Age: As a general rule, monthly publications have materials about 1–2 months old, and quarterlies about 3–5 months old.

Accuracy and Completeness: Fairly accurate, but rarely puts in data not obtained from individual or business members.

9.1.3 Government Reports

Audience: Government policy-makers and legislators, industry regulators, firms in a particular industry.

Purpose: Collect, analyze, and publish data, often highly aggregated, to inform government officials or to report on the impact of government policies.

Sources of underlying data: Usually uses data from its own collection efforts, such as surveys and interviews. Sometimes also use data collected and provided by trade associations and specialized research centers.

Level of Detail and Scope: Varies widely. Data is usually aggregated. Will not usually contain proprietary data.

Age: Tends to be dated, that is up to 12 months old, or old, that is over 12 months old, in most cases. When dealing with issues with high-level political or economic impacts, data may occasionally be less dated.

Accuracy and Completeness: Data collected directly from firms surveyed tends to be very accurate, but often, due to definitional issues, may not be directly comparable with other published data.

9.1.4 Government Records and Files

Audience: Agencies regulating the business/industry in question. Due to open records laws and regulations, much of this is also open to the general public.

Purpose: To allow agencies to conduct their business.

Sources of underlying data: Most often, these records contain materials provided by individual persons and firms to the agency or office.

Level of Detail and Scope: Tend to be highly detailed, as they are designed to provide data requested by the agency for the conduct of its business, to ensure compliance with rules and permits, etc.

Age: Vary widely. The contents can go back for decades in some cases. On the other hand, the frequency with which an agency updates its files depends on filing and reporting requirements, which can be from weekly to annual. The wait time for providing filings under FOIA requests can be significant.

Accuracy and Completeness: Materials, sometimes filed under oath, tend to be accurate. However, the scope of the materials can be very narrow, depending on the agency's filing and compliance requirements.

9.1.5 Security Analyst Reports

Audience: Investors in the securities of publicly traded enterprises. This ranges from individual to institutional investors.

Purpose: To provide information to clients and to demonstrate capability of research departments of the securities dealers, with the aim at improving its image, and thus its sales.

Sources of underlying data: Usually data comes directly from the firm being profiled. On occasion, key suppliers and distributors are interviewed. Industry-level data may be developed from firm-by-firm sources or from trade associations, as well as taken from government reports.

Level of Detail and Scope: Most reports tend to be focused in terms of time, covering events over the past few weeks or quarters. Some are annual. Detail varies widely, with most of detail coming from firm sources. It is heavily financial in orientation.

Age: Fairly current. Many reports will note the date of the previous report.

Accuracy and Completeness: SEC efforts are aimed at making these reports more objective and credible.

9.1.6 Academic Case Studies

Audience: Graduate level students in business, management, and related courses.

Purpose: To provide detail of specific business actions, set in a competitive context, so that students can prepare their own analyses of these cases.

Sources of underlying data: The targeted firms, including unpublished materials, studies by the professor writing the case, trade associations, and government agencies.

Level of Detail and Scope: Tend to focus either on history and long-strategy or on tactics in an unusual market and economic context.

Age: Usually 2+ years old.

Accuracy and Completeness: Very accurate, but often have a very narrow focus, or only a summary of the overall context.

9.1.7 Research Centers

Audience: Companies in the industry, government agencies involved in the industry, firms whose business is dependent on that industry.

Purpose: To develop new information, for sale, or to influence legislation and regulation, as well as to support academic research needs.

Sources of underlying data: Companies in the industry and government agencies, as well as independent research conducted by the centers.

Level of Detail and Scope: Tends to be very industry-centric, and is often heavily focused on strategy and policy. But can provide industry-level data not available elsewhere.

Age: Rarely current. It is usually dated or even very old.

Accuracy and Completeness: Except for advocacy pieces, they tend to be both accurate and complete. That is because accuracy and completeness is a part of the reputation the centers seek to develop.

9.1.8 Business Information Services

Audience: Companies extending credit to individuals or businesses; businesses selling market-level data.

Purpose: Allow customers to utilize the data in managing their businesses.

Sources of underlying data: For credit, creditors and the Target itself, as well as public records. For others, government surveys, as well as proprietary surveys.

Level of Detail and Scope: Credit-based sources tend to have a great deal of financial data from third parties. The degree of information on the individual business in non-credit categories tends to be significantly less. Surveys vary widely in scope and depth.

Age: Credit-based information tends to be current, with the exception of information provided by the Target. That may be old, or non-existent, depending on the cooperation of the Target. Survey-based resources tend to be dated to old.

Accuracy and Completeness: The accuracy of data provided by creditors and from public records tends to be very high. Data provided directly from the Target is often not crosschecked, and so must be considered as unconfirmed. Data from surveys, in the aggregate, tends to be accurate, but care must be taken to make sure that the survey encompasses all potential targets.

9.1.9 Advertising

Audience: Primarily existing and potential customers and consumers. However, advertisements in industry publications may also be aimed at developing or enhancing firm image.

Purpose: To sell product and services, either by featuring the product (or service) or by convincing potential customers and consumer that the advertiser is the right place to buy the product/service.

Sources of underlying data: The firm that places the advertisement is usually the sole source of data and photographs in the advertisement.

Level of Detail and Scope: Typically these are very low in detail and general in scope. However, sometimes photographs in advertising can provide data that the advertiser did not intend to provide.

Age: They tend to range from being current to being dated.

Accuracy and Completeness: Private trade regulation authorities tend to allow significant "puffing" so that their reliability can be questionable.

9.1.10 Want Ads/Job Listing

Audience: Potential employees of the Target.

Purpose: To obtain employees or executives to fill existing or anticipated needs.

Sources of underlying data: The advertiser alone is responsible for the details of the position provided in a want ad or home page job listing.

Level of Detail and Scope: While many want ads tend to be general, some online ads can provide a great deal of detail about the position to be filled, allowing inferences to be drawn about marketing tactics and strategy, as well as technology to be used.

Age: They range from being very current to being somewhat current. Most firms do not keep republishing a want ad, or keep it continually posted online, for very long after the position is filled.

Accuracy and Completeness: As the ad is selling the position, they tend to be incomplete. The details provided tend to be accurate, for reasons associated with equal employment opportunity and related laws.

9.1.11 Local Newspapers and Magazines

Audience: Individuals, and to a lesser extent business, living/working in the circulation area.

Purpose: Provide news and articles with a local focus, as well as providing a local "angle" on national stories. Also provides national wire services, columns etc.

Sources of underlying data: Most of local stories are developed by staff or local writers, but rely heavily on what they are told, except in "investigative" pieces.

Level of Detail and Scope: Most business articles are moderately detailed, but are not written for those in the particular industry. Sometimes, local businesses provide data to a local paper or magazine that they would not give to a national trade publication.

Age: Stories in the business section can range from the very current to those that are somewhat dated, while non-business stories tend to range from those that are very current to some that are only current. Some print stories may be updated on the publication's web site.

Accuracy and Completeness: These sources tend to be able to check most of the facts on which they rely, but sometimes they can be manipulated to generate disinformation.

9.1.12   Competitor's News Releases

Audience: A broad range of the firm's constituencies, including its shareholders, creditors, suppliers, employees, and customers.

Purpose: Keep constituencies informed; keep the firm name before the public and its customers; and build a positive image.

Sources of underlying data: Data in releases almost always comes directly from the firm, or from filings or reports it has already made public.

Level of Detail and Scope: Varies widely depending on the aim of the release. A release announcing a new officer will have less detail than one announcing the closing of a plant.

Age: They usually range from the very current to merely current. Some companies routinely purge their websites of old releases, while others archive them.

Accuracy and Completeness: While the material in the releases tends to be accurate, they are usually written by professionals to accomplish one or more particular ends. Often they disclose less than they conceal.

9.1.13   Competitor's Websites

Audience: A broad range of the firm's constituencies, including its shareholders, creditors, suppliers, employees, and customers.

Purpose: Keep constituencies informed; keep the firm name before the public and its customers; offer products/services; take orders; and build a positive image.

Sources of underlying data: Data almost always comes directly from the firm, or from filings or reports it has already made public. Many firms link the site to these other sources rather than reprint or report what is said there.

Level of Detail and Scope: Varies widely depending on the aim of each page of the site. Also, some pages are written by technical personnel, seeking to communicate with other technical personnel, while hiring opportunity pages may offer a look at the business' vision. Visuals, such as photographs, charts, or streaming media can contain significant detail.

Age: Ranges from very current to dated. This depends on how aggressive the firm's webmaster is in removing dated materials.

Accuracy and Completeness: While the material in the text tends to be accurate, all of it should be read in light of what that page is seeking to communicate and to whom. Often these websites disclose less than they conceal.

9.1.14 Directories and Reference Aids

Audience: Businesses, academics, reference centers, such as libraries.

Purpose: Collect and provide tabulated or cross-indexed data on a large number of firms or businesses that have something(s) in common.

Sources of underlying data: Almost always based on a combination of government sources, trade association data, and surveys voluntarily completed by the Targets themselves.

Level of Detail and Scope: Usually has significant detail, but within a very narrow spectrum.

Age: Most range from dated to old. Even web-based services may only be updated annually.

Accuracy and Completeness: Tend to be accurate, but, since they rely on voluntary surveys, completeness varies widely.

9.1.15 Technical Publications

Audience: Practitioners of the technology in question, both business and academic.

Purpose: Source of information on current technology, especially trends.

Sources of underlying data: Companies involved with the technology in question.

Level of Detail and Scope: While firms may cooperate, often that cooperation is conditioned on not allowing their names to be associated with particular elements of research and development.

Age: Current to Dated.

Accuracy and Completeness: Tend to be very accurate on technical matters, but show mixed results when dealing with firm- or even industry-level issues.

9.1.16 Catalogs

Audience: Customers and clients.

Purpose: Sell products/services. Enable customers/clients to understand Target's offerings.

Sources of underlying data: With the exception of "generic" catalogs, the Target provides all data. However, if the catalog includes products/services not developed by the Target, then descriptions and technical information almost always originate with source of product/service.

Level of Detail and Scope: Provides sufficient detail to allow potential customer/client to differentiate among competing products/services. May also, but not always, provide pricing or cost data.

Age: These are usually current to dated. Online catalogs may be current, but they also may be only an online version of an older print publication. However, a comparison of a current version with an older one can often provide interesting data on trends.

Accuracy and Completeness: Unless they are online, catalogs are incomplete due to lead-time issues. If the products/services originate elsewhere, the data is only as accurate and complete ads the ultimate provider.

9.1.17 General News Publications

Audience: Readers, usually subscribers, across the nation and economic spectrum.

Purpose: Provide current news, and, in the case of magazines, provide a context for that news.

Sources of underlying data: Usually have their own reporters and contributors. They, in turn, use a wide variety of sources, including but not limited to the Target.

Level of Detail and Scope: Level of detail tends to be less than in trade publications, but they are more likely to talk to experts and commentators to provide a sense of context and trends.

Age: Materials are usually fairly current. As a general rule, daily publications have materials 1–2 days old, weekly publications 1–2 weeks, and monthly publications, 1–2 months.

Accuracy and Completeness: While they tend to be fairly accurate, they are often less precise with critical details than are trade publications.

9.1.18 Books

Audience: Composed of general interest readers, along with those interested in a particular firm, industry or even executive.

Purpose: Some are historical studies; some are personal memoirs; some are expansions of feature stories or series in general interest publications.

Sources of underlying data: Vary widely. May include data not previously published or even disclosed elsewhere.

Level of Detail and Scope: Vary widely

Age: Rarely better than old (over 12 months in age), due to publication lead times.

Accuracy and Completeness: Depends on the author and the purpose of the work.

9.2 Primary Sources of Raw Data for CI

The availability, accessibility, accuracy, and level of detail of primary sources of raw data for CI will vary widely from industry to industry, from country to country, and from time to time. At times it will be almost impossible to obtain; at others, it will be freely available.

9.2.1 Your Own Employees

Sources of underlying data: Personal contact with competitors, customers, suppliers, as well as previous employment with them.

Level of Detail and Scope: Varies widely, but rarely very detailed. Gaining cooperation from employees, particularly from the sales force, may take significant effort.

Age: Data developed during previous employment tends to be dated, even if stated in the present tense. Other data tends to be as current at the contacts from which it is developed.

Accuracy and Completeness: Accuracy tends to be high but completeness varies significantly.

9.2.2 Industry Experts

Audience: You, your competitors, your investors, and those dependent on you and your competitors, such as suppliers and distributors.

Sources of underlying data: Vary widely, but generally include direct contact with the industry segment in question.

Level of Detail and Scope: Level of detail tends to be very significant. Focus can vary widely.

Age: Industry experts attempt to keep their data as current as possible.

Accuracy and Completeness: Tends to be very accurate and very complete.

9.2.3 Customers

Sources of underlying data: Direct experience with you and your product/service as well as direct contact with your competitors.

Level of Detail and Scope: Can be very detailed, particularly in terms of product/service characteristics and pricing, but rarely detailed with respect to strategic and technology issues.

Age: Current as of the time collected/experienced by the customer. But you must make allowance for the time between collection/experience and time that it is actually delivered to you.

Accuracy and Completeness: Can be very accurate, if customer is willing to share the data, and the data is current. Completeness is likely only with respect to marketing and product/service concerns.

9.2.4 Securities Analysts

Audience: Current and potential investors in the target firm.

Purpose: To develop and support customers of the investment firm itself.

Sources of underlying data: Target firms, including interviews, as well as industry experts, trade associations, and government data.

Level of Detail and Scope: Moderately detailed, and fairly broad in scope, including strategy.

Age: Most analysts produce reports at least quarterly, and sometimes more frequently.

Accuracy and Completeness: Data tend to be accurate and complete, particularly financial data.

9.2.5 Direct Contact with Competitors

Sources of underlying data: Competitor's own records and personnel.

Level of Detail and Scope: Can be very detailed and of very wide scope, depending on who is contacted, if this is a person-to-person contact.

Age: Can be extremely current.

Accuracy and Completeness: While direct contact should provide the most accurate data, care should be taken to make sure that the contact in fact has regular access to such data. In technical areas, it is not usually effective.

9.2.6 Suppliers

Sources of underlying data: Themselves, as well as competing firms that they serve.

Level of Detail and Scope: Level of detail can be very good, but the scope is most often highly restricted.

Age: Can be reasonably frequent, if supplier is in regular contact with competitors.

Accuracy and Completeness: If supplier will share competitor-sourced information, can be highly accurate and, within it limited scope, very complete.

9.2.7 Product/Service Purchases

Sources of underlying data: Reverse engineering of product/service, as well as sales, warranty, and service materials.

Level of Detail and Scope: Fairly detailed with respect to product/service details and performance. Narrow in scope, effectively providing support only for research and development as well as marketing-associated processes.

Age: May be dated with respect to other products/services mentioned in materials, as they reflect the situation as of the date prepared, which may be a long time before the product/service is purchased.

Accuracy and Completeness: Accurate and complete, but only as of the date prepared.

9.2.8 Trade Shows

Purpose: Provide opportunity for firms in the same market niche(s) to meet potential customers, distributors, suppliers, etc.

Sources of underlying data: Direct contact with attendees, as well as materials provided to attendees.

Level of Detail and Scope: Can be somewhat detailed, particularly with respect to technical and product/service characteristics. Also, direct contact may elicit significant gems of data on intentions in the marketplace in the near term.

Age: Very current.

Accuracy and Completeness: Can be very accurate and often quite complete.

9.2.9 Industry Conferences

Purpose: Multi-firm meetings to deal with current and future developments facing a particular industry or market niche.

Sources of underlying data: Speakers at the conference, attendees, and exhibitors, i.e., vendors, if any.

Level of Detail and Scope: Can be quite detailed, and scope is usually broad. These are rarely useful for getting data on marketing issues.

Age: Reasonably current. However, the speakers usually prepare their presentations when invited, so that these may be somewhat dated.

Accuracy and Completeness: Can be very accurate and complete.

9.2.10 Technical/Professional Meetings

Purpose: Bring together technical/professional employees for continuing education in their common discipline and technology.

Sources of underlying data: Presentations made at the meeting by speakers, direct contact with attendees, and exhibitors (vendors).

Level of Detail and Scope: Can be quite detailed, and scope is usually not very broad. However, they are rarely useful for data on marketing or strategy issues.

Age: Reasonably current. However, the speakers usually prepare their presentations when invited, so that these may be somewhat dated.

Accuracy and Completeness: Can be very accurate and complete, if very focused.

9.2.11 Facility Tours

Purpose: Public relations effort by facility owner to familiarize the public with its business.

Sources of underlying data: Actual viewing of working facility or virtual online tour. Rarely is contact with employees at work permitted.

Level of Detail and Scope: Very general in terms of information. Close examination is usually not permitted. The result of that technical information, while potentially accessible through direct observation, can only be acquired if the person taking the tour is a trained observer.

Age: Current.

Accuracy and Completeness: Most, but not all, tours are highly controlled. However, employees can become so inured to the presence of a tour that they may inadvertently discuss business matters in presence of the tour.

9.2.12 Retailers and Distributors

Sources of underlying data: Customers, competitors and sales representatives.

Level of Detail and Scope: Can be very detailed, particularly in terms of product/service characteristics and pricing, but rarely detailed with respect to technology issues.

Age: Varies. When retailers and distributors collect data, it is often dated when they get it.

Accuracy and Completeness: Can be very accurate, if retailer or distributor is willing to share the data. Completeness is likely only with respect to marketing and product/service concerns or general changes in strategy, as expressed to these sources.

9.2.13 *Advertising Agencies*

Purpose: Collect data from clients and develop marketing campaigns as a part of service relationship.

Sources of underlying data: Direct contact with firm and media.

Level of Detail and Scope: Very deep in the marketing arena and, reasonably broad scope, except for technical matters. Often may have access to client intentions or strategy.

Age: Very current.

Accuracy and Completeness: Very accurate and complete, at least with respect to products/services represented.

9.2.14 *TV, Radio, Internet Programs, Interviews, and Webinars*

Purpose: Provide business information and news to readers, viewers, and listeners.

Sources of underlying data: Interviews with key officers of target firms, sometimes accompanied by analysis from industry experts and/or securities analysts.

Level of Detail and Scope: Often very high-level, but, if carefully reviewed, can provide insights into the way in which executive(s) view the competitive environment and how they plan to deal with it.

Age: May not always be very current. Print and recorded interviews may be done one or more weeks before release.

Accuracy and Completeness: Very accurate, but not necessarily complete. That often depends on how informed and aggressive the interviewer is.

9.2.15 *Speeches*

Purpose: Vary widely. May be promotional, defensive, or even just made for personal reasons.

Sources of underlying data: Speaker and those assisting in preparing remarks, as well as preparing for questions and answers.

Level of Detail and Scope: Often very high-level, but can provide insights into the way in which executive(s) view the competitive environment and how they plan to deal with it.

Age: May not always be very current. Text may be prepared weeks before presentation.

Accuracy and Completeness: Very accurate, but not necessarily complete.

9.2.16 Internet Chat Groups and Blogs

Purpose: Communication among individuals with common interests. These may be as investors, retirees, potential investors, consumers with complaints, etc.

Sources of underlying data: Personal experiences of those on the list or sending messages to the site.

Level of Detail and Scope: Usually not very deep. Scope generally limited to subject(s) of the list itself. Those that are complaint or feedback sites may include data on how firms actually handle complaints as well.

Age: Very current, but sometimes old data, even misinformation, can resurface.

Accuracy and Completeness: Very hard to gauge. Many members conceal identities under Internet handles.

Chapter 10
Who Else Can Help You?

At this point, you have completed your own hands-on research, and have worked the members of your network. Where else can you go—assuming always you have the time, and budget? There are three options:

- An in-house CI unit,
- Other in-house research-oriented teams, and
- Outside CI vendors.

These resources can be a fast and rewarding way to conduct, or at least supplement, your own CI research. For example, individuals and teams scattered throughout your own firm can quickly answer the following kinds of questions—provided you approach them properly (more about that later):

- What are current customers saying about your competitors' products/services?
- Why are they switching (or at least what do they say about that)?
- What are customers' purchasing agents saying to your sales representatives about your competitor's new product and service offerings?
- What past industry conferences have your firm's employees attended, what did they learn there, who else attended, and where are the materials that they took back?
- Who in your firm has paid outside contractors and services for one-time reports and studies? On what topics? Where are those reports? What do they say?
- What product and service introductions or rumors have production personnel heard from your suppliers?
- Who are your current competitors? On what basis are they identified as such? Who has a copy of their current price lists and catalogs?
- What do others in your firm know about competitive products/services? Do they have samples? Where are they? Have they been reverse engineered? By whom? With what results?
- Who provides your firm with what goods and services? Who are the appropriate contact people at each enterprise that you can contact for more current data and leads to further interviews?

10.1 Strengths and Weaknesses

Now, we will quickly go over the relative strengths and weaknesses of each of these three options as compared with your own work.

10.1.1 Doing it Yourself

10.1.1.1 Strengths

As we explained earlier, one of the great benefits of doing the CI you need yourself is the fact that you are the client. By that, we mean you should be able to determine exactly what your needs are, exactly what the extent of your current knowledge is, and what kind of an end product you need in order to support your next decision or action. It also has the additional benefit of putting you in the way of the data gathering, so that you can understand the importance of some aberrational fact or other anomaly. Involving anyone else might mean losing the early warning that this provides to you.

If you think that we are coming on too strong, you should realize that the very top executives, the C-suite, are already doing something like this. A 2009 Forbes study disclosed that

> [m]embers of the C-suite search for information themselves. While delegating research may be part of the stereotype of a C-level executive, it is not the reality. More than half of C-level respondents said they prefer to locate information themselves, making them more self-sufficient in their information gathering than non-C-suite executives (Forbes 2009).

10.1.1.2 Weaknesses

As we have pointed out, the weaknesses of doing yourself are pretty obvious: this is yet another demand on your already limited time; it requires you to step back and evaluate a problem differently from ways you have done before; and you have no one to cross check your research or analysis before you conclude it.

10.1.1.3 Which is Best When?

Doing it yourself is probably best in most situations. For example, if you have an internal competitive intelligence unit which you can utilize, taking a few steps into the research and analysis before you contact it will enable you to sharpen your request and enable that unit to focus more efficiently on getting you the needed intelligence. When dealing with other research units, you should understand exactly what you need from them in the way of data and analysis to support your competitive intelligence work. It is pointless to go to these groups and merely ask for "help" when you do not know what you have and what you need.

In terms of dealing with outside contractors, you need to understand what you need to know and perhaps as importantly, what you do not want researched. The more you can refine and define your needs and what you know, the better a third-party can assist you. However, do not make the mistake of simply telling a third-party to go out and do the work and report back 6 weeks later. For a project which takes a significant amount of time you should always build in a reporting and review process.

10.1.2 Getting Help from an Internal CI Unit

10.1.2.1 Strengths

One of the strengths of an internal competitive intelligence unit is, of course, that the unit is trained in providing CI. A second strength lies in its combined knowledge of your industry as well as your own firm.

10.1.2.2 Weaknesses

One of the key weaknesses of such a unit, at least with respect to you, is that its mission may be to serve others. If it is tasked to serve you, then you should consider approaching it, as we describe below, as a best practices customer.

10.1.2.3 Which is Best When?

When they are available to you, you should strongly consider using in-house CI resources even before using other internal research resources and outside contractors. As we note below, you should be doing certain things to make sure that you are the best customer possible for a CI unit so that it can provide the best results to you.

10.1.3 Getting Help from Other In-house Research Resources

10.1.3.1 Strengths

By other in-house research resources, we mean units such as marketing research, consumer awareness, and other research-oriented units. The strengths of these units are that they are already embedded in your firm and have access to their own regular sources of data.

10.1.3.2 Weaknesses

The key weaknesses of these units, at least with respect to providing you with
competitive intelligence, is that their approach is going to be largely quantitative,
while your approach, needing competitive intelligence, is largely qualitative.

10.1.3.3 Which is Best When?

These internal research resources can be useful when you are seeking supple-
mentary information that may well been developed by others in your firm for other
purposes.

10.1.4 Getting Help from External CI Suppliers

10.1.4.1 Strengths

By external suppliers, we mean outside CI contractors, to whom you outsource
some or all of your competitive intelligence work. Among the key strengths of
such providers is a range and depth of experience in collecting, analyzing competitive intelligence for a variety of industries and in a variety of countries.

10.1.4.2 Weaknesses

A key weakness in using an external supplier is that you must dedicate time and
resources to bringing it up to speed. It may only be necessary to bring it up to
speed on a particular project if you're dealing with the CI supplier with which your
firm has worked in the past. If you are dealing with a supplier with which your firm
has not worked in the past, you may have to spend additional time and resources
educating this outside person or group.

10.1.4.3 Which is Best When?

Using outside CI firms is probably best when you are facing "hump" projects, that
is, ones where you do not have enough time to be able to generate the needed end
product. In addition, they are particularly useful in situations where the research
calls for direct or indirect contact with competitors and other sensitive targets. This
is not to say they should be used to get around any restriction on contacting
competitors; that is wrong. Rather, if your firm's ethical standards allow this, they
can become involved in an elicitation effort against these targets quickly and
efficiently.

10.2 How Can You be a Better Customer for Intelligence or Data?

Whether you are going to an in-house CI team that is working for you or your unit, asking for help from an internal competitive intelligence team that supports some other function, contacting your market research specialist, or using an external competitive intelligence research and analysis firm, you have to think of yourself as a client or as a customer. As we have noted before, articulating what you want and then beginning to approach it is not something to be done lightly. It must be done in a disciplined manner. Dealing with these resources is similarly disciplined, but in a different way.

To get the best from these resources and people, you have to help them and let them help you. We have identified nine key areas where you should prepare yourself and learn to work with a third-party provider of intelligence or data, whether inside or outside.

10.2.1 Need to Know Versus Want to Know

The first key is to determine whether you want to know something, or whether you need to know something. The former can be classified as something just above curiosity, while the latter as something near necessity. The distinction is very important. In some firms, the CI units have the right to decline to take an assignment, that is, to refuse to work on it. One of the most common reasons that they exercise this authority is that there is no clear need that the CI team can see for the intelligence being sought.

You should clearly communicate the difference to any third-party provider. Tell them that you "need" the intelligence, and avoid using the word "want".

10.2.2 Start Early

As we mentioned earlier which talking about you doing your own research, timing is a major issue. We all know that, all too often, we never have enough time to do something right. The same is even truer with someone that you ask to help you, for they have to begin and finish their work in sufficient time for you to make use of the data or intelligence he/she can provide you. In addition, the more time that any third-party provider has, the more time he/she can spend on good analysis, following on sound data locations and development. In other words, if you can give more time, you can get a better product. In addition, by giving a third-party provider enough time, their costs, or your costs, or both, can be held down.

In addition, you cannot assume these sources are always available to you instantaneously. The CI team may have been co-opted in connection with a possible acquisition, tying up all of their time for the next 3 weeks. A third-party vendor may have a conflict of interest, or its industry expert may be away for 3 days at a conference. Always pay attention to timing, and always be realistic. If you need an interim report because you have to give an interim report, tell a third-party provider right at the beginning. Do not wait to tell them after it has begun its work. The way they do their work, if there is an interim report, may differ from the way they would do their work if they are only delivering a final product. Every time you change the parameters, their costs and your costs change, and they never change for the better.

10.2.3 Do not Assume What is Available

When dealing with any third-party provider, particularly in an area with which you do not have much familiarity, never assume what is available. One of the biggest problems that CI analysts have with their internal and external customers is that very often what the customer, in particular an inexperienced customer, brings to them is request not only for intelligence, but very particular data, possibly gathered in a very particular way.

For example, what if you ask a third-party provider to get you the script of a business-to-business telemarketing call campaign of a competitor, as a client once did to us? You may find that the provider tell you that this is virtually impossible, simply because even the callers themselves may not have the full script. On the other hand, if you explain that you want to know what the focus of the calls is, and to what kind of firms they make calls, ask that. Do not assume that the only way to do this is to look at a script; provide the target, and not the tactics.

10.2.4 Watch Personal/Institutional Biases

As we mentioned, you have to be careful about your own biases, as well as any institutional biases at your own firm. And here, as elsewhere, by biases we mean those factors that control how you process data and the conclusions you draw. Biases are not necessarily bad—but they are always present. If you recognize them, or at least admit that you have them, that acknowledgement can help a third-party provider immeasurably.

For example, you may have a personal belief that a particular competitor is not much of a threat, but you're asking a third-party provider to help you by reviewing its marketing strategy. If it is important to know its marketing strategy, do not bias the research by indicating that you do not expect to find any strategy, or at least any meaningful strategy. Now you are transmitting your bias to a third person.

Institutional biases can also infect a third-party provider. For example, it may be "conventional wisdom" within your firm that customer loyalty is the most important factor in protecting and gaining market share for your business. And that may well be true. But do not assume that, and do not burden the third-party provider with the assumption that the competitor or competitors you have targeted must necessarily operate in the same way. While the odds of it being correct are probably fairly high, by starting with such a bias, you will almost automatically cut off any research that might contradict that bias.

10.2.5 Where did You Fail Before? When were You Surprised?

When you are trying to guide third-party research, think about the past. Ask yourself, in a similar situation where did you or your firm fail before? Ask, were we surprised, and if so when? The present is not necessarily same as the past, but by articulating how past issues were handled, or mishandled, you are already providing an insight into what you or someone else missed in the past and what should be explored more carefully.

In addition, taking this kind of an approach allows you to step back and develop a slightly different perspective, which as we mentioned elsewhere, is particularly useful in designing and then managing competitive intelligence research and analysis.

10.2.6 Identify Decisions that You Can Make When You Get the Intelligence

When you are seeking competitive intelligence from a third-party provider, it should be for a reason. Most often that reason is to provide input into a decision that you must make, or one or more steps you must take. It can be particularly helpful to a third-party provider to know what decisions are at issue, or what steps you are considering taking.

The importance lies not only in improving their ability to understand your needs, but also that you are harnessing their skills and experience during their research and analysis. For example, we worked on an assignment where the client was seeking to determine the investment strategies of a number of its competitors, large and small, public and private. The reason for the research was that the client was seeking to expand and wanted to know whether or not some of its competitors might be looking to expand in the same geographic areas.

That slight difference in emphasis meant all the difference in the project. Not only were we looking for information on investment strategies, we were also looking for any other information that might be useful to the client in making its decision about expansion. In this case, we identified one small private player in this market, which was already in the targeted geographic market. A slight degree of additional digging disclosed that it might be available for purchase. Following delivery of the report, the client purchased that firm. Knowing the decision that was to be supported made all the difference between a nice report on investment strategies, and one which enabled it to take a particularly important and decisive action.

10.2.7 Be Realistic

When asking a third-party for assistance, be realistic. For example, if you ask for report by the end of this month, do not expect that if you then put aside for 4 months, it will still be accurate. The rule of half-life applies in competitive intelligence as well as it does with radioactivity: every intelligence analysis has a half-life, that is, a period of time after which it has lost half of its value. It may be 1 day, 1 week, 1 month, or perhaps as long as 1 year. But that happens. Tell the third-party provider when you expect to use it as well as how you plan to use it.

If you do not use the intelligence when you expected to, do not expect that it will get better with age. Be prepared to ask for some assistance in updating it particularly if the report indicates something on the order that the competitor being targeted has yet to make a decision on the subject. By the time you get around to using report, the competitor may well have decided.

10.2.8 When do You Need it and in What Form?

Be very clear not only as to what you need, but in what form. If you need a PowerPoint presentation because that is how your firm communicates, ask for help in doing that. But make sure that you have received all of the intelligence, say in a written report, so that you can describe what it is when you make the presentation. In other words, you want to get deliverables that permit you to master the facts completely to present your case.

Also, be honest with your third-party provider as to the level of detail you need. Do you need approximations, such as a 15–20% growth rate? Or do you need more precise data? Always keep in mind that the more you ask for, the more it costs in terms of time and money.

If you have to have a document that is going to be circulated, tell the third-party provider that. Why? The provider may include things in a document that you should not circulate to others. The most common is the identification of individuals, by name, with whom the third-party provider has spoken. Now, you may want this information to assure yourself that elicited interviews were conducted with the right individuals. However, you do not want to put any of these resources at risk. For example, if your third-party provider talks to one of your own employees that previously worked at a competitor, do you really want that person's name circulated widely? We can assure you that that person does not want that to happen. Not that they are ashamed of what they did, but they likely do not want to be cast as disloyal to a former employer.

10.2.9 Demanding/Working with Pushback

The last key is to learn to work with pushback. Pushback is one of the benefits that you do receive when working with a third-party.

Pushback means that the individual talking to you, as an end-user, about your competitive intelligence needs does not merely listen to what you say, write it down, and arrange to deliver it in 3 weeks. Pushback means that the individual asks, and re-asks, questions about what you want, why you want it, in what form you want it, when you want it, and what you will do with it. In other words, he/she is probing to make sure that what is delivered completely meets your needs. Some managers and executives tend to be taken aback on this, believing that it somehow a challenge to their ability to define the problem. Nothing could be further from the truth.

Pushback means that the third-party is treating your request as a serious, well-thought-out request. What he/she is probing for is to develop an understanding of not only what you need, but what you do not need, to provide not only a better work product for you, but one which is done on a more cost and time effective basis. Not until you have a long working relationship with a third-party provider should a provider feel comfortable about diminishing the amount of pushback.

10.3 Where Can I Go Inside My Own Firm and How Can I Get Help?

You can consider this as an advanced version of the networking we discussed earlier. What you are doing is contacting others within your firm, and seeking their help in completing your own competitive intelligence research. There are three key points to keep in mind here:

- You are seeking their help.
- You cannot order them to help.
- Help is not a one-way transaction.

It is very difficult to go in "cold" to any one of these groups and just ask for help on little or no notice. It is much preferable that you spend time in advance introducing yourself, telling them what kind of work you will be doing, and indicating that you would appreciate their assistance at some point in the future. You must be prepared to, and in fact be willing to, provide individuals from each of these groups with help should they come to you for assistance.

You are asking people to do things in addition to everything else that they do. We all know that everyone is overworked, so you must be respectful of their time. While *you* may have a deadline of 10 working days from today that does not mean that they will be able to assist you in the time that you have. When asking for help, take what you can get. Do not push. Frankly, you have nowhere to push from and you will only create dislike among people that could potentially help you.

Have your needs clearly and sharply defined. These people are going to try to help you; it is not their job to formulate your assignment or your research for you. It is up to you to articulate, as precisely as possible, what you want them to help you with. For example, if you are talking to people in sales, don't say "I need some information about our competitor's sales team". Salespeople are not in the business of being news reporters—they are in the business of selling. Hone your questions down to the point where you can ask something like, "How many in-house salespeople do our competitors have and what firms do they use to supplement their sales team?" It is very likely that these people can help you on that immediately.

Be respectful. When you are done with your work, always thank those who helped you in some way. As we said before, and we cannot over-stress it, an e-mail or short call thanking him/her for help is always more than appropriate—it is really necessary. In addition, if someone has given you particularly important assistance, feel free to put in a good word for them with a direct supervisor. Of course make sure that this is okay with them first. They may be helping you "on the side." And do not want their supervisor to know that they have enough extra time to help you.

10.3.1 Sales

Your sales team can be incredibly valuable because it is your firm's direct contact with your customers. That means the sales team is on the front lines. And not only do they know about your customers, they often run into information about your competitors given to them by their customers when, for example, the customers are bargaining with the competitors.

But remember the sales force's job is to sell. Approaching them for a quick answer to a well-thought-out question is the best way to go. The most valuable data you can collect from them is that which they know off the top of their heads. Do not expect them to go out and do any kind of research for you. Every minute they spend researching for you is a minute less making sales calls. Do not expect them to do the impossible—they have enough to do right now with too little time.

10.3.2 Marketing and Market Research

Your marketing department and your market research team are research-oriented. This has the advantage of putting them in the middle of raw data with some marketing intelligence of their own.

You must appreciate that most marketing and marketing research departments are overwhelmed with projects and deadlines. If you are in the retail business, you will already be aware of the pressure that the end of the month reports from SymphonyIRI or Nielsen puts on them. Always respect their schedule.

Also, realize that most marketing and market research departments are heavily quantitatively oriented rather than qualitatively oriented, or to put it more simply they tend to be made up of numbers people. Your best bet is to ask for something that they probably already have, but are using for a different purpose. For example, you may be looking at the way a competitor approaches a particular market. It is unlikely that the marketing department will have statistics on this, but individuals who have been with the department for sufficiently long time may have an opinion on this. Take that opinion. Always confirm it, of course, but again take what is offered and do not request, much less demand, that they manipulate their data for you.

If your research has disclosed a new item, or new event, that the marketing or market research department would-be interested in, unless there are severe constraints on discussing it, feel free to share the insight with your contact and marketing, and make sure that you indicate that you appreciate their help in developing this.

10.3.3 Research and Development

Your R&D staff functions in the world of science and engineering. That does not mean that it has failed to develop insight on your competitors. Rather, it means R&D has access to different information from different channels.

When working with R&D, make sure that you understand the technical terms in the areas in which you are inquiring. Few things frustrate someone who was scientifically trained as much as individuals who misuse or downright mangle scientific terms and acronyms. Master them before you talk.

Your best use of people in R&D is to extract from them a perspective. Explain your needs precisely and ask highly focused questions. Having said that, always keep the conversation open enough that they can respond to your area of interest, even if they do not have an answer to your question. In other words, encourage them to talk, using perhaps some of the elicitation techniques we described earlier on telephone interviews.

A real case may illustrate what we mean. Several years ago, we worked with a large industrial firm which was attempting to reach out to many different departments to determine how these departments might help a competitive intelligence unit that was in the process of being built. This is something that those in the CI profession refer to as "intelligence auditing". During the audit, a member the future CI team talked with the head of R&D and asked what associations the R&D employees were members of, where they could go to annual meetings, and perhaps report back information.

Rather than simply answering the question the head of R&D asked. "Why do you want to know?" The response was, "Well we might want to have you find out whether a particular competitor is getting ready to develop a nonmetallic product for this market space or whether or not they have considered developing it and then abandoned it." The head of R&D responded "We know they've given up doing it. So?" After metaphorically picking himself off the floor, the would-be CI team member asked "How do you know that?" The head of R&D responded that he could tell that simply by the resumes he got from researchers at the competitor, researchers who were leaving because a major project in this area was being shut down. The would-be CI team member asked, "Why didn't you tell us?" The response was "I did not know that you are interested."

10.3.4 Information Center

The Information Center (IC), to previous generations known as the corporate library, can be a strong ally for you. The job of an IC employee is to provide answers to questions. What you should remember is that information they will provide to you is virtually all secondary, and can range over a broad period of time. So, before you begin asking for assistance here, think about restricting your questions. For example, do you want any information that is older than 2 years? Do you want it from outside the United States or on a global basis? Do you want information from business reporting sources or from sources such as patent databases or technology reviews?

Unless you focus, a good researcher in the IC can literally overwhelm you with data. And having too much data is almost as bad as having none. The difference is that when you have no data, you realize you do not know the answer, but when you have too much data you live in fear that somewhere in the middle of those 150 screens of material lies the answer to your question, but you have not found it yet.

It is very useful to introduce yourself to individuals in the IC and explain how you would like to use their services. Be open to the way that they want to receive request for services. This is not only a matter of back billing, time allocation, and the like; it is also a matter of professional courtesy. Individuals in the IC have worked hard in the past to be trained for this area, and occasionally may feel somewhat intimidated by people that are doing competitive intelligence research, which includes primary, elicitation, interviews.

Some ICs index reports purchased by individuals throughout your firm. A brief check may lead you to a valuable report—and to someone you may want to talk to about his/her interest in the subject.

10.3.5 Call Center

The Call Center is not an area with which you will probably work very much, but you should be acquainted with it. Visualize the call center as a series of nerves reaching out. The Call Center is dealing on a minute-by-minute basis with customers, former customers, and potential customers. The staff has a job to do, in a context where courtesy and attention to the customer is invaluable. However, it is their contact with the customer that is critical to you.

Unless your firm's policies are clearly to the contrary, you should never try to approach an employee in the call center. Approach a supervisor. What you are looking for are threads of information. For example, when dealing with returns, is there any indication that the returns are prompted by new, aggressive marketing or pricing campaign by a competitor? It is unlikely that the individuals in the call center will be thinking about these questions without prompting. However, you may be able to convince a supervisor to chat with a couple of people to get their impressions. These impressions can be very valuable, but do not take these impressions to be hard fact. Validate them yourself, unless you can convince the call center supervisor to include such surveillance and reporting in the call center dialog.

10.3.6 Consumer Insights

Consumer insights is a newer area, often encompassing parts of marketing, elements of the call center relationship, and even some marketing research. Its focus is customer facing, and it is designed to help a firm understand its customers and consumers better. In so doing, these individuals may well not only receive, but in fact affirmatively gather, competitively useful data. In a few firms, a consumer insights department is where the competitive intelligence activities are now focused.

This can cause a problem just as with a free-standing competitive intelligence unit in the sense that some individuals do not take well to someone else trying to "do their job". If you are going to deal with members of a consumer insights team, approach them diplomatically, offer future assistance, and ask for help, if possible, in that order. Diplomacy above all pays dividends here.

As with the individuals in marketing, a new perspective you develop that may be useful for them is something that you should share.

Reference

Forbes (2009) The rise of the digital C-suite: How executives locate and filter business information, Forbes/insights. http://www.forbes.com/forbesinsights/digital_csuite/index.html. Accessed 12 Oct 2011

Chapter 11
How Do You Make Sense of Your Data?

11.1 An Overview of Analysis

Now, you will usually be facing a mass of data, data which has come to you in no particular order. It is most likely incomplete, and contains misinformation and even disinformation, in addition to useful information. You must draw actionable conclusions from what seems (and often is) an unstructured mass of raw, unevaluated data.

Analysis is the process by which you handle this mass of data, so you can produce a solution to your problem. Remember, that usually your problem is typically made up of several parts, so you face the problem of solving each part. In turn, having solved all of the parts, you should then be led to a solution of the whole problem.

One view of intelligence analysis is that it is made up of four separate sub-processes:

- Amassing
- Incubation
- Enlightenment and
- Corroboration.

11.1.1 Amassing

First of all, amassing involves the collection of raw data directly related to your problem. But it is broader than this. It really includes all of the data touching on the subjects that you have been accumulating, both within this business context, and over a lifetime of experience.

J. J. McGonagle and C. M. Vella, *Proactive Intelligence*,
DOI: 10.1007/978-1-4471-2742-0_11, © Springer-Verlag London 2012

11.1.2 Assimilation

Here you carefully review all of the raw data. This is not the same as a one-time through reading. It entails a slow and careful reading of all materials, usually two or more times. You should consider reviewing the materials in different ways—once by source, another time by topic, and another time in chronological order. The goal is to master the materials, even those which do not, at first, seem to be important. Experience shows that it is often those bits of raw data which, after progressing through the next steps, can have unanticipated importance.

11.1.3 Incubation

Here, think about what you have collected. Both consciously and unconsciously, you will begin to assemble the facts in various ways, so that one or more logical pictures begin to emerge. This aspect overlaps with assimilation because it starts as the data begins to come in, and it involves both an evaluation of the validity of the data, as well as the credibility of the source. It also marks the beginning of interpretation of the raw data.

11.1.4 Enlightenment

After a long study of your problem, and of the data bearing on it, the real meaning of the data and a solution to the problem at hand presents itself to you. Sometimes this enlightenment actually can occur in a flash, what we in CI call the "Ah hah moment." More often, it is a gradual process, combing for little signs that add up to something big. Enlightenment encompasses your actual analysis of the raw data, in the sense of drawing conclusions, and also covers data interpretation, as well as having you form and test hypotheses.

11.1.5 Corroboration

Here you seek to prove, or disprove, the solution that comes out of the enlightenment process. This is not the same as merely verifying the raw data you collected. Rather, it involves having you draw conclusions, and testing the validity of your conclusions against the observed facts, before using them in your decision-making.

From here on, we are going to give you hands-on tips on how to do all of this in dealing with proactive competitive intelligence.

11.2 Pay Attention to Everything You Found

For example, having identified your present competitors, you want to focus on what they have going for them. How do you do that? Review the CI you have acquired about each competitor as a means to answering the following questions: "What is this company's single greatest competitive asset?"

Here are a few real examples of how paying attention small bits of data can add up:

- One major corporation annually listed its major areas of activity at the back of its annual report. That by itself did not provide much data about its strengths and weaknesses. However, a comparison of these reports and other promotional documents over time showed one line of activity was progressively moved further and further down the list. This reflected an unannounced corporate decision to de-emphasize that activity over time. For your firm, that may mean this activity will be less of a future competitive threat.
- A call for capability materials from one company was met with the reply that there were no materials of that kind now available, but that they would be available in about four to six weeks. The reason given for this was that these materials were being substantially revised as a part of a new marketing approach. So this temporary unavailability was a hint of some upcoming change in marketing direction. To you, this could be a warning to avoid committing your firm until you know the new direction. Alternatively, it may mean that you, as one of its competitors, have a temporary window of opportunity to go after new business while the target company is at a temporary disadvantage.
- A call was made to one division of a company using the phone number found on the home page. However, a second call was required, to another, newer number, to make contact. That number, assigned to a new office location, was a tell-tale warning of an as yet unannounced divisional restructuring.

11.3 Analyzing Your CI Data

Analyzing the results of proactive intelligence research is a different process for each project because the analysis is a function of the task, the data collected, and your growing experience. A number of tips, based on how CI professionals do it, can help you to manage each assignment:

- Organize your preliminary research results to help you conduct your analysis.
- Be imaginative and alert to the importance of detail.
- Identify and eliminate disinformation.
- Locate patterns and determine their significance.
- Always think about the accuracy of the data and the reliability of the source from which you got it.

- Seek out anomalies and understand why they occurred.
- Try generating vital data using disaggregation.
- Be sensitive to data that should be there but is not.
- Be sensitive to data that should not be there but is.
- Do not make assumptions—find out.
- Review your results to ensure completeness and consistency.
- Mentally separate your data from your conclusions.
- Keep in mind that your goal is always to convert raw data into proactive intelligence.

Some of these points need no further elaboration, so we will limit our comments to a couple of them.

11.3.1 Organizing Your Research Results

11.3.1.1 Outlines

How do you assemble your data to help you produce your analysis? One way is to use an outline in your word processing software.

- First, create an outline of the topic you are analyzing.
- Then, as you review the raw data, insert a piece of data as many times as needed under *every* appropriate heading.
- Repeat this for all of your data.
- Then read it, section by section. What conclusions do you see? What problems? What gaps?

11.3.1.2 Data Versus Conclusions

Another tactic is to separate your data, visually or physically, from your preliminary conclusions. Then go back through it and make sure you have enough data to support each of your conclusions. If you do not, add the data you did not include in the right place or just drop the conclusion as unsupported. Also verify that you have drawn conclusions from all data you first assembled. Otherwise, why did you include a piece of data in the first place?

11.3.1.3 Sources

Typically, your raw data can be classified in three ways:

- What your competitor say about itself.
- What one competitor says about another of its competitors, including your firm.
- What third parties say about your competition.

When you reorganize your data by area of coverage, such as dividing it among categories such as costs, markets, consumer relations, and financial strengths, you may find that data in one category is dominated by one of these three categories of sources.

If that is the situation, decide what that means and whether it is significant. For example, if virtually all of the information about a competitor's manufacturing costs comes from third parties who are not also competitors, that fact may tell you something about the quality of the data.

- Among the potential conclusions could be that the cost data is not likely to be accurate, because you can find no way in which these outsiders could have derived it.
- Alternatively, it could mean that the outsiders were provided data by your competitors, so that you have data that is quite accurate.
- To determine which alternative is more likely, you will have to immerse yourself in the data and put yourself in the place of both the outsider and your competitor.

11.3.1.4 Accuracy and Reliability

Once you have gathered your raw data, never assume that all of that data is accurate. You must establish the probable accuracy of that data for it to have any meaning at all to you. Lacking that foundation, you may be basing your intelligence analysis on a concoction of very good data, marginally correct data, bad data, and even disinformation.

Even if you find that you can draw some sort of conclusion from that mishmash, you risk drawing the wrong conclusion. However, once you have a sense of the relative accuracy of the individual pieces of data, you can then begin to analyze all of your data properly.

For example, suppose that you have located an article about your target company in a publication whose articles have been quite reliable in the past. However, later, you find that this data appears to be contradicted by a report whose source you generally consider to be unreliable. Given these two facts, you can reasonably conclude that the article is likely to be correct. The fact that it was contradicted by an unreliable source may actually make it all the more likely to be correct. Doing this, you can then minimize the impact of the other, less reliable, data in your analysis, or even ignore it completely. In fact, a conflicting account from an unreliable source may make you more secure in your reliance on a relatively reliable data source.

As this example shows, in evaluating raw data, you are actually engaged in several different tasks. You have to try to establish the reliability of the source of the data that you have. To do that, you should try and determine exactly where the data was produced, because the identity of the originating source may help you to estimate the reliability of the data.

You also have to estimate the accuracy of the data itself. Sometimes you do this by comparisons. For example, is one piece of data confirmed by data from an independent source? If so, it may be accurate. Before you jump to that conclusion, you should complete other steps, such as spotting and then eliminating false confirmations of data.

There are three basic steps in evaluating the accuracy of raw data:

1. Identify the actual source of the data so you can evaluate the reliability of the source.
2. Estimate the data's accuracy so you can classify the data.
3. Eliminate false confirmations.

Establishing the Reliability of the Source

"Reliability" means asking yourself how credible is the source of your data. You estimate how much you can believe any data coming from a particular source based on its past performance. On the other hand, "accuracy" relates to the correctness of the particular piece of data you have. There, you are estimating how correct the data is, based on factors such as whether it is confirmed by data from a reliable source as well as the reliability of the original source of the data.

To evaluate the probably accuracy of any data, you must have a sense of its ultimate source. That means figure out why the data was produced, collected, and released, a process we covered earlier. You have to understand the origin and history of each piece of data, because those facts are critical to its analysis.

You can safely assume that all data is produced and released for some certain purpose. The key is to figure out for what purpose. For example, if a national trade association releases market share and growth rate data on its members, it may have collected those facts for a hearing before the US International Trade Commission (USITC). The association may have tried to convince the USITC about the potentially severe impact of certain imports on the US economy. So the data presented to the USITC by the association would support this position. It will tend to be presented in a light that advances the association's case. This does not mean it is "wrong." It only means that it was collected and packaged to make a certain point.

Taking this example further, assume that each of the association's members that provided data for the USITC hearing also produces its own market share and growth rate estimates in its separate annual report to shareholders. These estimates, when totaled, differ from the association's estimates. Is one wrong? No.

The two estimates are different because they were produced for different purposes and two different audiences. To use either or both, you first have to understand the origins of the data. Then you can decide which you should use, if either. So, when you begin to analyze raw data, keep in mind the need to tie your evaluation of the raw data to the reason it was created, not to the reason you are seeking it.

You must not ignore the origins of data, either. Data is only as good as its source. With the earlier example, the industry data you got from the files of the USITC or in one of its reports does not have as its real source the USITC. That is just where you found it. Instead, the data's real source should be considered to be the member companies that provided the raw data, as well as the association itself, if the organization performed any analyses of the data from its members.

Keep in mind this informal rule: Unless you can establish otherwise, assume that every place from which you get data has its own point of view that permeates any data from that source.

Estimating the Data's Accuracy

One way to look at intelligence is to visualize a jigsaw puzzle. You start with many individual pieces of data that initially seem to be meaningless and unconnected. When you put them together correctly, however, they produce a picture—valuable Proactive intelligence. One key difference between competitive intelligence and a jigsaw puzzle is that, when you deal with intelligence, you may not have all of the pieces, and you may even have to remove some of the pieces you have and not use them.

Eliminating False Confirmations

Confirming data serves to assess its accuracy. In the long run, it also serves to assess the reliability of its source. A false confirmation is a situation in which one source appears to confirm data you obtained from another source. In fact, there is no real confirmation, because the first source may have obtained its data from the second source, or they both may have received it from a third source.

For example, consider trying to confirm data from a trade or professional publication. These narrowly focused magazines and newspapers offer both benefits and disadvantages as data sources. On the plus side, the fact that these publications have a specialized focus allows them to present stories with a degree of detail that is almost impossible to find in the more general media. On the other hand, many trade publications depend heavily on press releases issued by companies in their industry, often printing them without any editing or fact checking. That means that seeing the same report in two trade publications may only mean that both have printed the text of the same press release. Again, this results in a false confirmation, due to the fact that both publications have a common source for their data.

Precision

Be careful that you understand exactly what has been said and what has not been said. Increasingly, many businesses and industries use their own private languages. Sometimes, this provides clarity and precision for those insiders and other participants who use common data. In other cases, it only serves to keep outsiders from understanding what is actually going on.

Assess the Consistency of Data

Merely because you have consistent data does not mean that you can immediately draw a conclusion based on that data. When your research seems to provide consistent estimates, it can mean one of several things:

- The data and your conclusions are valid.
- No one ever questions this particular "revealed truth", so you have no idea if this consistency means the data is accurate.
- All the data has a common source, so there is merely a false confirmation.

Look at the data in question and analyze it, keeping all of these possibilities in mind. Make sure you know why data is consistent before you rely on it.

Patterns

Always start by looking for direct indications of what you are seeking. For example, in a high technology business, is your competitor hiring more researchers, building more new facilities, being awarded more patents, or devoting more funds to R & D than in the past? In practice, you should not really expect to find such directness. What you will generally find is data that involves indirect hints.

In that case, it is important for you to identify patterns and determine their significance. Pattern recognition is critical. Looking again at an earlier example, reading the annual report of a corporation for one year may disclose that a particular operation is a separate division. But reading these same reports covering a period of years might reveal that the prominence with which the results of that division are reported has increased substantially, and that the head of this division seems to have a direct track to the CEO suite. This may, in turn, reflect an increase in the relative importance of that division to the parent corporation.

Omissions and Displacements

Do not be afraid of finding that you lack data on something. The presence of a gap may be significant, or it may simply represent an area where more work needs to be done. In particular, the presence of a "gap" should alert you of the need to develop or supplement that data.

What is not present after you have finished your research can often be as significant as what is present. For example, you may find that a competitor is planning to sell a particular operation. From previous analysis, you may have found that this operation is a highly profitable one. If you can find no reason for the proposed sale, you should consider that a significant omission. Then try to establish what the most plausible reasons might be for this possible action.

Consider this real case:

- A major corporation was concerned about the strength of a competitor that operated two separate, but related, businesses. The first was in the consumer sector; the second was a raw materials division that provided some inputs for the consumer sector. The consumer business seemed strong, but raw materials division seemed weak.
- However, a closer look indicated that each division had the same promotion track and equal representation in senior management. Comparing this company with others similarly structured ultimately revealed that this company did not price inter-company transactions on the same basis as others. The result was that the consumer division was made to appear more profitable and the raw material division less profitable than they actually were. But internally, both divisions were treated as equals.

Check for Anomalies

What is an anomaly? It is when data does not fit. It is usually an indication that your working assumptions are wrong or that an as yet unknown factor is affecting results.

Something out of the ordinary should not be automatically rejected as an aberration or even a mistake. It may just be an anomaly. If you spot an anomaly, first ensure that it is not the result of a mistake in the way data was presented or collected, such as transposed numbers or a misquote. If it is not a mistake in that sense, look for other data that indicates that this is something which is true or could be true in the future.

What you are doing is actually attacking your own assumptions by using the anomaly to test them. The existence of an anomaly may indicate that your basic assumptions about what is true or what is possible are not correct.

A classic example of an anomaly and the potentially revealing conclusions drawn from it can be seen in a case involving some "futures" researchers. One such group was reported to have correctly predicted significant and imminent engineering developments based on an apparently anomalous remark by President Reagan in a 1980s State of the Union message. That remark concerned development of a jet that would fly at 15 times the speed of sound or faster. According to these researchers, conventional jet fuel could not be used for aircraft flights over 5 times the speed of sound. That remark, coupled with information from technical journals, led them to the conclusion that the US Government was developing a hydrogen fuel for jet flight. The anomaly was that the President was talking about developing a plane for which there was no fuel. As the State of the Union Address is a carefully prepared, written document reviewed by many government advisers, the researchers conclude that, as the remark was in the written text, it could not have been a casual slip. (Browne 1988; Federation of American Scientists 2011; Page 2007)

Keeping alert for anomalies has another benefit. By doing it, you help prevent yourself from falling into a common trap: the predisposition to subconsciously reject a deviation from a known trend or situation until a new trend or situation has been conclusively established.

Is All of Your Data Relevant?

Finally, eliminate data that is not relevant to your specific intelligence needs. It is almost always the situation that you will bring in large amounts of valuable raw data through your effective collection efforts. However, typically, these same efforts also harvest large amounts of "good" data that is not directly relevant to your problem. You have to put irrelevant data aside once you see that it does not directly pertain to your problem.

Preconceptions

When analyzing raw data, as the final step toward producing proactive intelligence, you must be very careful about your own view of the world and of the particular problem which you bring to the task.

Given that you are who (and what) you are, how can you protect against this form of blindness? Start all your analysis with as little reference to your expected outcome as possible. Only in approaching it that way will you spend the proper focus on understanding facts.

Drawing Inferences

When you study raw data and try to come to a conclusion about what it all means, you are drawing inferences. That process involves coming to a conclusion in light of both logic and of your own past experience. However, that same process also may cause you to shoehorn incoming data into your own preexisting beliefs or to see what you expect to be there. In other words, your own experience acts as a screen on the data as well as an aid in analyzing that data.

Just being aware of the difficulty of dealing with inferences can help you avoid its problems. Here is a brief test to see whether you are having problems dealing with inferences. Ask yourself, as each new piece of data comes in, which of these is your reaction:

- "That particular fact is incorrect." or "That fact is correct."
- "That fact must be incorrect."

If your response is the second one, you may be fitting the data into your preexisting beliefs, instead of testing it to see what it really means.

Drawing Conclusions

Your ultimate goal is to draw a conclusion. That conclusion should be logical. But your conclusion may not seem to be logical to you every time. If, for example, you are trying to determine what a competitor will do under certain circumstances, your ultimate goal is to anticipate how that competitor thinks. That is, in turn, based on its track record, its corporate culture, and how it perceives its competitive environment. To you that perception may not be careful and realistic, but that is not the concern here. The issue is not whether the target is correct; it is what the target actually perceives.

11.4 Develop Perspective for Your Analysis

You may properly ask if "Haven't I been doing analysis while I was collecting data"? The answer is, in part, "Yes". Every time you decide to pursue a resource, to interview a particular individual, or to move on to another direction, you have engaged in some analysis. But that analysis is focused on how valuable or useful is what you have found to your quest and where you should go next. That is not the same as analyzing the entire body of your data.

Now you have finished your research. By that we mean, you have (1) run out of time (2) run out of budget, and/or (3) are having individuals whom you interview or try to interview referring you back either to secondary resources or to individuals that you have previously tried to interview or interviewed ("closing the loop").

Now, you must break the mindset that you developed during the data collection. You collected data in a certain order, which is almost certainly neither chronological nor topical. The individuals you have interviewed were available on their schedule, not in an order chosen by you. So try one or more of these five techniques before you start determining what it is you actually think you have found:

1. Arrange your information and notes chronologically, that is by what period of time they deal with. Now go back and read them, both from the oldest to newest and then the newest to the oldest. What gaps do you see? If you have time, consider conducting supplemental research to address them.
2. Arrange your notes and information by topic, such as by sales and marketing, production capabilities, personalities, etc. Once again look for gaps in your information. Again see if supplemental research is valuable or possible.

3. Put aside your research materials. Start by writing what you think you know about the target, based on what you have knew before you started (if you have time, try doing this at the start). Put that down even if you believe that your preconceptions, for that are what they are, have been contradicted or undercut by your research.
4. Write out or just outline, whichever you prefer, what you have learned from your research. Clearly identify those things that are certain, that is confirmed facts, from those that are your conclusions or just inferences. This is not to say that your conclusions and inferences are not important; they are, and can be correct. But you owe it to yourself to separate out what you knew, from what you know, from what you think you know.
5. Take a look at your starting questions and then write up answers grouped by the type of resource, that is government, or interviews, or trade association materials, or even particular interviewee. Do you find that one or more of the questions you are answering are really dependent on data from only one source? If so, spend your last little bit of time trying to support or attack that information on your own.

What you are trying to do is to stand back from your research and look at it through someone else's eyes. Now, since you do not have someone working with you, the next best thing is to try to break through your own preconceptions, adopt a perspective different from your normal perspective, or use some other device to force you to stand back and think openly.

You should do this even though you are the consumer of your own research product. In fact, because you are the end-user of the research, it is important that you force yourself to step back to break preconceptions, to change perspective, to be open to new views, and to challenge what you think you know, than it would be in a normal competitive intelligence situation. There, a competitive intelligence specialist reports to an end-user. That automatically provides for at least two perspectives on the research. When you are combining the roles of the researcher, the analyst, and the customer, you must take extra care to make sure that you do not jump to any conclusions, much less blind yourself to important anomalies that you should pay attention to.

If you will have to circulate something based on this research and perhaps even defend it to others, then you will be facing the same issues that CI professionals face at this stage. There is a tremendous amount of good literature on making presentations and communicating, in the CI literature, in government intelligence literature, and in related areas, from presenting scientific papers, to making legal arguments in court. If you need, or think you need assistance in this area, seek it out and learn from it.

If you are eventually going write up a document, even for your own use, there is a drafting trick which you can use. It requires that you have been somewhat disciplined in recording your research somewhere, rather than relying on your memory to retain everything. This is a critical point and should not be overlooked. Do not keep everything in your mind—keep notes as you go.

For example, if you are struggling to make sense of a problem, try doing a *reverse*: chronologic listing of what you know and what you suspect. If you start with the most current and then move backwards in time, this will force you to think about each piece in turn and not just sort through these facts.

There are literally dozens of analytic tools you can use in competitive intelligence. Consider starting with the tools with which you are most familiar and learn to master them. For example, the SWOT (strength-weakness-opportunity-threat) table is very often misused. Even if you use it regularly, refer to one of the resources we have listed at the end of the book to make sure that you really do understand how to use it and what its shortcomings are. More sophisticated tools, such as the analysis of competing hypotheses, are probably not necessary to achieve what you need. They are designed for professional analysts to make sense of a vast array of data, often collected by many assistants.

Most of the time you will not need particularly sophisticated analytical tools. However, you should be open to using tools which you have not used in the past, for the same reasons that you must have your mind open to facts and sources that you are not used to seeing. Learning about a new analytical tool, even if the result is that you decide you do not use it, teaches you to look at problems differently. If you do not try to continually educate yourself about analysis, you will put yourself in the situation of the poor individual described by Mark Twain: "To a man with a hammer, everything looks like a nail." Make sure you have more than a hammer in your toolbox of analytical techniques.

The issue of follow-on communication is not as critical for you as it is for those who are full-time CI professionals because, in most if not all cases, you are the end-user of the intelligence. That means you do not need to struggle with issues of whether or not to use PowerPoint, how to express probabilities in a consistent way, how properly to graphically display data, and other issues which are of real concern to those who provide finished intelligence to others.

This does not mean that when you finished your research you are done. For your own use and for others who may need to know more about what you found, you should commit your research to some kind of document, whether it is a chart, a PowerPoint presentation, outline, or memo. If you think that you may have to come back to it, generate a listing of sources, in general terms; putting it at the end of the document may be useful. Be careful about connecting the names of individuals you interviewed with your finished intelligence—they may not be comfortable about being contacted later by someone who got their name from your report.

11.5 Select the Right Tools

As we mentioned, there are many, many tools you can use to analyze your data. But, you should select from among them only *after* you take a look at your data and tentative conclusions. Do not start your research by assuming that you will be putting together a Boston Consulting Group matrix or a SWOT chart. If you do that, you may subconsciously be trying to find the facts and conclusions that fit into that particular analytical model; rather find a model which helps you deal best with the facts that you have. For additional information on what tools are available to you, see the books listed in Chap. 13.

Most Common CI Analysis Tools and Techniques

Asset turns
Benchmarking: traditional
Benchmarking: competitive and shadow
Blindspot analysis
Cash flow and conversion analysis
Competitor positioning analysis
Cost analysis
Country and area risk analysis
Credit/debt analysis
Crisis management assessment
Critical success factors
Cultural assessment
Distribution strategy analysis
Environmental assessment
Financial ratio and statement analysis
Forecasting
Game theory
Gap analysis
Individual assessment/profiling
Industry analysis
Legislative and regulatory analysis
Management profiling
Market analysis
Modeling and simulations
Organization assessment
Patent mapping
Political and economic assessment
Porter's "Five Forces" Model
Portfolio analysis
Ratio analyses
Relationship mapping
Response modeling
Reverse engineering
Scenario development and analysis
Shadowing
Share/growth (BCG) matrices
Statistical and econometric analysis
Structural and trend analyses

Supply chain management analysis
Sustainable growth analysis
SWOT (strength, weakness, opportunity, threat) analysis
Technology forecasting
Trend analysis and projections
Value chain analysis
War gaming

11.5.1 What Tools Will You Use Most Often?

Of course, you do not need to use all of these tools, or even to master all of them. In fact, a study by SCIP found the following about CI professionals:

> Competitive intelligence professionals can apply many analytical techniques to turn information into actionable intelligence. But CI practitioners generally prefer to use only a few techniques, and those preferences have not changed much over the years. In this survey, CI practitioners indicated they use two analytical techniques (competitor analysis and SWOT) frequently and others occasionally. (Fehringer et al. 2006, p 9)

To those two, you should add war gaming, which is growing in popularity among CI practitioners.

11.5.1.1 SWOT Analysis

SWOT analysis is a strategic planning method used to assess the Strengths, Weaknesses, Opportunities, and Threats (SWOT) involved in a project. It involves specifying the objective of the project and then identifying both the internal and external factors that are support or impede obtaining that objective. SWOT's four considerations are

Strengths: those characteristics of the firm that give it an advantage over others in the industry.
Weaknesses: those characteristics that place the firm at a disadvantage relative to others in the industry.
Opportunities: the external chances to make greater sales or profits in the competitive environment.
Threats: the external elements in the environment that could cause trouble for the firm.

It is essential to identify the SWOTs because the rest of the process depends on fitting them together and then relating them properly. Then you can determine whether or not the specified objective can be attained.

The SWOT analysis template is normally presented as a grid, comprising four sections, one for each of the SWOT headings: Strengths, Weaknesses, Opportunities, and Threats. You first list the key elements of SWOT on the outside. On the inside, you combine them and derive an observation based on the combination. Table 11.1 deals with a decision about expansion or contraction of production by a cereal company.

11.5.2 Competitor Analysis

Competitor analysis involves an assessment of the strengths and of the weaknesses of current and potential competitors. This aims at bringing all of the relevant sources of competitive analysis into one framework to support effective strategy creation, execution, monitoring, and adjustment.

11.5.2.1 Competitor Arrays

One useful technique of competitor analysis is to build a competitor array. The steps are as follows:

- Define your industry, in particular the scope and nature of the industry.
- Determine who your competitors actually are. You can also do this for potential competitors.
- Determine who your customers are and what benefits they expect.
- Determine what the key success factors in your industry are.
- Then rank the key success factors by giving each one a weighting where the total of all the weighting add up to one (1.0).
- Rate each competitor or potential competitor on each key success factor. This usually involves a 1–10 scale
- Multiply each cell in the matrix by the factor weighting.

See the following example (Table 11.2)

In this example, Competitor #1 is rated higher than Competitor #2 on national distribution (9 of 10, compared to 6 of 10) and economies of scale (6 of 10). Competitor #2 is rated higher on name recognition (8 of 10). Here Competitor #2 is rated slightly higher than Competitor #1 (26 out of 40 compared to 23 out of 40). When the success factors are weighted according to their importance, Competitor #1 draws closer to Competitor #2.

Additional columns can be added to expand this analysis. For example, in one column, you can rate your own firm on each of the key success factors to see how you measure up against the competition. In another column you can list the ideal standards of comparisons on each of the factors. These are benchmarks, reflecting the operation of a firm using all the industry's best practices. That way you can

Table 11.1 SWOT analysis

	Strengths 1. Current profit ratio increased 2. Employee morale high 3. Market share has increased	*Weaknesses* 1. Legal suits not resolved 2. Plant capacity is 3. Lack of strategic management system
Opportunities 1. Western European firms looking for investments 2. Rising health consciousness in selecting foods 3. Demand for cereals increasing annually	Opportunity-Strength (OS) Strategies Sell non-cereal food subsidiaries in Europe (S1, S3, O1) Develop new healthier cereals (S2, O2)	Opportunity-Weakness (OW) Strategies Develop new cereal-based snack products to replace low margin cereals (W1, O2, O3)
Threats 1. Unstable value of dollar 2. Current packaging is not "green"	Threat-Strength (TS) Strategies Develop new biodegradable containers (S1, T2)	Threat-Weakness (TW) Strategies Close unprofitable non-US operations (W3, T1)

(Adapted from US Department of Health and Human Services 2011)

Table 11.2 Table to show competitor analysis

Key success factor	Success factor weight	Competitor #1		Competitor #2	
		Rating	Weighted score	Rating	Weighted score
National distribution	0.4	9	3.6	6	2.4
Name recognition	0.2	5	1.0	8	1.6
Economies of scale	0.1	6	0.6	5	0.5
New product development	0.3	3	0.9	7	2.1
Totals	1.0	23	6.1	26	6.6

compare your firm and all of your competitors with the best attainable performance.

11.5.2.2 Competitor Profiling

Another approach is to use the checklist at the end of Chap. 7 to develop a formal competitor profile. The power of competitor profiling is multi-faceted:

- These profiles can reveal strategic weaknesses in rivals that your firm may exploit.
- The proactive nature of competitor profiling will allow your firm to anticipate the response of rivals to your firm's planned strategies, the strategies of other competing firms, as well as to changes in the competitive environment.

This kind of proactive knowledge will make your firm more agile. With current intelligence on competitors at hand, you can implement your offensive strategy more quickly to exploit opportunities and capitalize on your own strengths. Similarly, you can deploy a defensive more precisely to counter the threat of rival firms exploiting your firm's weaknesses.

11.5.3 War Gaming

The rationale for running a business war game is that it can prepare you and your firm to identify and then deal with changes, potential and actual, in the competitive environment. It does this by allowing you to consider, proactively, how different players, that is, competitors, customers, suppliers, etc., can react to the change, and to each other.

Business war games of this type are most useful in testing, developing, and refining strategy at the SBU level, as well as at the product, project, and geographic market levels.

There are a wide variety of types of war games, including multi-player, computer simulations, as well as those that involve the application of game theory.

A less celestial variation uses CI to develop in-depth profiles of competitors and other key players, allowing you and your employees to role play your actions and responses to a variety of situations. The purpose is to enable you to be able to predict and then respond to moves of competitors and others, testing your firm's strategy in a fairly realistic setting.

Consider how this war game played out, relying on CI to support it:

> A food processing company, a large player in the ready to eat (RTE) market faced growing pressures from competitors...and a threat to its leadership position. The business unit management decided to run a war game to determine whether the source of its presumed competitive advantage [technology] remains valid and sustainable in light of changes in the RTE industry.
>
> [The game's results showed that the] company suffered from a pricing disadvantage against its largest rival, and its belief in its **technology lead was a blindspot**.... Analysis of the company competitive advantage focused on its leading position in healthy foods, but also a need to strengthen this image which was lapsing slowly....
>
> The outcome: The Company embarked on "nutrition first" long term strategy.... Over...5 years it regained profitability and leadership position.... (Gilad 2011)

11.6 Allow Enough Time

As we indicated earlier, you should allow plenty of time for your analysis. A government manual dealing with Open Source Intelligence, which is very close to what you are doing, recommends that 25% of the time involved in creating a report should be taken up with analysis (North Atlantic Treaty Organization 2001). Since you are not "producing" a report, this actually comes out to 33% of your time being involved in establishing your needs, doing your research, and analysis should be involved with analysis.

References

Browne MW (1988). Clean hydrogen beckons aviation engineers, New York Times, 24 May, 1988, www.nytimes.com/1988/05/24/science/clean-hydrogen-beckons-aviation-engineers. html?pagewanted=all&src=pm, Access date 13 Oct 2011

Federation of American Scientists (2011). X-30 National Aerospace Plane (NASP), http://www.fas.org/irp/mystery/nasp.htm, Access date 13 Oct 2011

Fehringer D, Hohhof B, Johnson T (eds) (2006) State of the art: competitive intelligence. Society of Competitive Intelligence Professionals, Alexandria

Gilad B (2011) Market leader under pressure. http://www.bengilad.com/corporate_strategy_simulations.php. Access date 4 Oct 2011

North Atlantic Treaty Organization. Nov (2001). NATO open source intelligence handbook. http://www.au.af.mil/au/awc/awcgate/nato/osint_hdbk.pdf. Access date 11 Oct 2011

Page L (2007). Hypersonic plane project confirmed by DARPA, http://www.theregister.co.uk/2007/09/19/blackswift_hypersonic_confirmed_uprated_sr71_blackbird/, Access date 13 Oct 2011

US Department of Health and Human Services, Administration for Children and Families (n/d). SWOTanalysis: strengths, weaknesses, opportunities, and threats, http://eclkc.ohs.acf.hhs. gov/hslc/tta-sys-tem/operations/Management%20and%20Administration/Planning/Program% 20Planning/2SWOTAnalysisS.htm, Access date 12 Sept 2011

Chapter 12
Let us Review: Doing it the First Time

Well, now that you have decided to do this yourself, and we have given you all of this help, how do you do it the very first time? To begin with, why not write an outline? Actually, write down questions. Put down the first question, the really big question that you want to answer. Then determine the three or four specific questions that have to be answered so that you can draw the conclusions for the big question. If any one of those is complex, just break it down again. Now you have an outline of your research. As a matter of fact, a good trick is to save this outline on your computer and write up your report, if you are giving it to someone else or just retaining it for your future reference, by answering each question. Then, simply turn the questions into statements and use them as headings in a longer report.

When framing your questions, think in terms of action. That is, what decision will you make or what action will you undertake as a result of your findings? If you cannot identify a decision or action, ask yourself why you are spending the time and other resources to dig all of this data out. While "good to know" or "just getting some background" is not necessarily a waste of your time, you will eventually learn that in many cases, it is exactly that. If you need to do background research, do background research. But that background research should have as its focus supporting a decision, or even the framing of a decision, and eventually the taking of or avoiding action.

When you have identified what you are looking for, try to work out how long you have to get the answers to your questions. This does not mean building a timeline, but be aware of the fact that if you are researching this for a meeting, whatever you have when the meeting starts is what you must present.

Do not feel that you must achieve perfection. You will not be able to do that. Intelligence is not like research in higher education, where the premium is put on perfection and closing all of the open questions. Intelligence is designed to be actionable, and timely. In other words, a perfect answer to an important question delivered a day late may be worthless, but a largely correct analysis delivered on time may be invaluable.

J. J. McGonagle and C. M. Vella, *Proactive Intelligence*,
DOI: 10.1007/978-1-4471-2742-0_12, © Springer-Verlag London 2012

Let us say that you have about two weeks to do this. Set a reminder to yourself at the end of the first week to do a review of exactly what you have accomplished. Then allow plenty of time for your analysis. Realistically, you are doing analysis as you go along, but always allow time to review your research and dig for the additional insights which you may have missed. While there are no hard and fast rules on the subject, experience and research indicates that, given a choice between spending 90% of your time on data collection and only 10% on analysis versus 50% on data collection and 50% on analysis, in most cases you will be better off doing the latter. So leave plenty of time for analysis and for any last-minute additional data gathering.

Now what how do you conduct your CI research?

If an article in a trade publication triggered your research, start there. With respect to the article, read it again. Is anyone quoted in the article? When was the article published? From there determine when the reporter or news service had to have gotten the information in the first place. You now have an idea of how old that information was. Use the article as kickoff point: if you can contact someone who was quoted in the article, even if her/his contribution was peripheral to your concern, consider starting with that person—call them, or e-mail them. What you are looking for is not just information from them, but also a lead to somewhere, something, or someone else. If you cannot call an interviewee, can you contact the reporter? Realize that if you do talk to a reporter, particularly one covering your industry, you have to exercise great care. As a reporter is not likely to give up information without an exchange, now or in the future, from you, exercise great discretion in making such contacts.

If your source for the information that led you on your research chase was a report from one of your sales representatives, contact that sales representative. Was there anything else that she heard, saw, or even suspected? Again you are looking for leads beyond her. If she mentioned that she heard this information from a store manager, ask if she will do you the favor of going back to that store manager. Remember, you are simultaneously conducting research and trying to build up and keep a network of good contacts, so always be prepared to help out those who help you. We covered the issue of building a network earlier, but keep in mind that each time someone helps you, thank them, even if the help was not very significant, and add them immediately to your network.

As you go along, first look at each piece of data that you collect as being an element in building up an answer to your question. Also treat it as a way to find other sources of data. For example, if you find a reference to your problem in a trade publication which is not one which usually covers your industry, research further in that publication. They may have covered this particular topic, target, or problem earlier. Never be afraid to ask someone who says they "cannot help you" to suggest someone who might be able to do so. If they do (1) thank them, (2) get as much contact information as you can, and (3) ask if you can use his name when you reach out to the new source. If you can, it often means the difference between getting cooperation and getting very little.

Here are several tips which experienced CI researchers use in managing the course of their research:

- Make sure you at least tried to access some resource in each of the large groups that we have given you. It is not that each group will always have the data you are looking for, but without digging into each one, you cannot be sure that somehow a particularly useful source of a bit of data has escaped you.
- Stop about halfway through your research. Look at, either physically or in your mind, where you have gone. Have you found that interviews have been much more useful than secondary research? Has what you been looking for been disclosed, at least in part, by filings with federal, state, or local governments? Use these insights to reset the balance of your research.
- Look at your research sources. One effective rule is that if you keep coming back to the same resources, such as particular agencies, individuals, public records, and trade associations you have probably exhausted most of the sources that can help you. It is time to shut down the research and move into the analysis.

For those of you who are going to be writing up a document, there are a number of drafting tricks which you can use. Each of them requires that you have been somewhat disciplined in recording your research somewhere, rather than relying on your memory to retain everything. This is a critical point and should not be overlooked. *Do not keep everything in your mind—always keep notes.*

For example, if you are struggling to make sense of a problem, try doing a chronologic listing of what you know and what you suspect. If you start with the most current and then move backwards in time, this will automatically force you to think about each piece in turn rather than skim over the entire batch.

There are literally dozens of analytic tools you can use in competitive intelligence, which we listed earlier. Consider the tools with which you are most familiar and learn to master them. Even if you use a particular tool regularly, refer to one of the books we have listed at the end of this book to make sure that you really do understand how to use it and what its shortcomings are.

This does not mean that when you finished your research you are done. For your own use and for others who may need to know more about what you found, you should commit your research to some kind of document, whether it is a chart, a PowerPoint presentation, outline, or memo. If you think that you may have to come back to it, adding a listing of sources in general terms at the end may be particularly useful. Be careful about using names of individuals you interviewed—they may not be comfortable about being contacted by someone who got his/her name from your report.

Chapter 13
What's Next?

13.1 For You

13.1.1 A Career Advancer

As we have said, engaging in proactive Competitive Intelligence (CI) is a way to help you do your job, whatever that is, better. But, there is more. Remember, we urged you to look at the world differently to be able to develop your own CI. Now, once you are used to doing CI, you should find that you look at the world differently. What can that mean for you?

- Well, you will be less likely to blindly believe any "conventional wisdom", at least without testing it.
- You will have a broader and deeper knowledge about competitors. And you will also have a better sense of what you do not know. Over time, you find out just how valuable that can be.
- You should be able to look more objectively at your own firm. You will understand better what it is doing—and not doing.
- Now, think about how you can use these new-found skills and perspectives.
- Look at your own firm—objectively. Where are the hot spots, the growth areas, the next great places to be? Conversely, where are the dead ends, the units most likely to be downsized, sold or closed? Where are you? Where should you be? Now, get there.
- How does your firm match up against your competitors? How does your unit match up against its opposite number? Is it decision time?
- Now, look at your industry? What threats are facing it? What opportunities do you see coming? Where should you be?
- What is your job? Does using CI add more value to what you are doing? Do others know and appreciate that? Where else can you use your now-enhanced skill set? How?

J. J. McGonagle and C. M. Vella, *Proactive Intelligence*,
DOI: 10.1007/978-1-4471-2742-0_13, © Springer-Verlag London 2012

- Does your new set of skills mean you are capable of another, different career, or does it mean you should get more training on CI or in other skills?

Remember, you are in a world that continually changes. Unless you are willing to be proactive, you will be only reactive.

13.1.2 Continuing Your Education

13.1.2.1 Books

The following books can be very useful for your self-education. We specifically have avoided including very technical or esoteric sources. If you need them, these books will help lead you to them. While each book covers many topics, we have indicated where we think they will help you the most (Table 13.1).

13.1.2.2 Additional CI Education

Check the annual conventions and chapter meetings of associations to which you already belong to see if they have any sessions on CI. In the past several years, for example, the following US groups have all had such sessions:

- American Marketing Association–marketingpower.com/
- Association of Independent Information Professionals–aiip.org/
- Association for Strategic Planning–strategyplus.org/
- Product Development and Management Association–pdma.org/
- Professional Pricing Society–pricingsociety.com/
- Society for Insurance Research–sirnet.org/
- Special Libraries Association–sla.org/
- Strategic and Competitive Intelligence Professionals–scip.org/

Finally, keep your eyes open for free webinars that can improve your new-found skills. For example, one business networking site has periodically offered a free webinar on "cold calling". That could be a very useful source of tips on setting up telephone interviews and then eliciting information.

13.1.2.3 Training

You can enroll in structured public courses regularly offered around the world by two private firms:

- Fuld-Gilad-Herring Academy of Competitive Intelligence–academyci.com/
- Institute for Competitive Intelligence–institute-for-competitive-intelligence.com/

Table 13.1   Books for reference by subject area

Title	Collecting data	Data analysis	Using CI
Bensoussan BE, Fleisher CS (2008) Analysis without paralysis: 10 tools to make better strategic decisions FT Press, Upper Saddle River		◆	
Clark RM (2nd ed. 2007) Intelligence analysis: a target-centric Approach. CQ Press, Washington		◆	
Clauser J (rev. and edited by Jan Goldman) (2008) An introduction to intelligence research and analysis. The Scarecrow Press, Lanhan	◆	◆	
Duncan R (2006) Competitive intelligence: fast, cheap and ethical. AuthorHouse Books, Bloomington	◆		
Fahey L (1999) Competitors: outwitting, outmaneuvering, and outperforming. Wiley, Inc., NY.			◆
Fleisher CS, Bensoussan BE (2003) Strategic and competitive analysis. Prentice Hall, Upper Saddle River		◆	
Fuld LM (2006) The secret language of competitive intelligence: how to see through and stay ahead of business disruptions, rumors and smoke screens. Crown Business, NY.	◆		
Gilad B (2004) Early warning: using competitive intelligence to anticipate market shifts, control risk, and create powerful strategies. AMACOM, NY.		◆	
Kahaner L (1996) Competitive intelligence: from black ops to boardrooms—how businesses gather, analyze, and use information to succeed in the global marketplace. simon and schuster, NY.	◆		
Miller JP (ed) (2000) Millennium intelligence: understanding and conducting competitive intelligence in the digital age. CyberAge Books, Medford.		◆	◆
Mintz AP (ed) (2002) Web of deception: misinformation on the internet. CyberAge Books, Medford		◆	
Nolan J (1999) Confidential: uncover your competitors' top business secrets legally and quickly— and protect your own. HarperBusiness, NY.	◆		
Walle AH III (2001) Qualitative research in intelligence and marketing: the new strategic convergence. Quorum Book, Westport		◆	
Waters TJ (2010) Hyperperformance: using competitive intelligence for better strategy and execution. jossey-Bass, San Francisco			◆

13.2 Defending Against Your Competitors' CI Efforts

Now that you are comfortable with doing CI on your own, you will quickly find that you are very sensitive to how much CI your competitors may be collecting on your firm. Here are some basic principles involved in protecting you and your own firm from the competitive intelligence efforts of your competitors:

- Think ahead: Have you considered what the shrinking size of camera phones and digital recorders means to you? Do you really want any nonemployees attending your new product launch to be able to snap a photo of a confidential overhead or record the brand new presentation and email it in a matter of seconds?
- Do not give away sensitive competitive information to everyone: For example, you can mention in a press release or on your web page that a new product is launching in the summer, but you do not have to reveal the exact launch date or where it will be sold first. Also, use nondisclosure agreements with partners, suppliers and consultants working with you on key projects—expansions, research and IT projects among them. The goal is to have them keep quiet until you are past the critical ramp-up times. For particularly sensitive data, remind them to brief their own employees–again.
- Know who you are talking to: Who are the people at the other end of the telephone call, exactly what are they doing, and why are they asking you questions about your firm? Just because the call seems to come from inside, that only means it was transferred to you, and does not mean that you are talking with a co-employee. Why not Google them while they are on the telephone *before* you start talking?
- Examine your own Web site: Are you revealing too much? Remember, just because a page is not indexed for public access does not mean others won't find it. They will search for it and they will find it!
- Find out who is talking about you: Do an Internet search under your firm's name, and see what comes up. Then, run a back-link search. What sites did you find back linked to your own site? Why? Are current or former employees who are posting resumes on job-search sites or on business networking sites revealing competitively sensitive data?
- Use common sense: Do not carry or display materials, like tote bags or caps, with the logos or names of unannounced products/services in public places, like airports. Do not talk about business on your cell phone in the middle of a crowd. And do not work on sensitive documents on your laptop or tablet while flying— you never know who your seatmate works for.
- Do not over file: Keep to an absolute minimum the competitively sensitive data you file with government agencies, such as the US Securities and Exchange Commission, the US Environmental Protection Agency and its state equivalents, as well as with local zoning and planning commissions. If you have to provide sensitive data, provide it in a separate document or PDF file, label *every* page as

"proprietary and confidential" (or "trade secret", if it really is one), and formally request in writing that it be kept separate from the public file, if not permanently, at least for a pre-determined time.

- Share the effort: If you have corporate security personnel or have legal staff focusing on intellectual property issues, such as trademarks and patents, make them all aware of what you are doing to protect your part of the business from the CI efforts of your competitors. They may help you to identify what data you should protect and for how long.
- Educate others: Explain to all of your co-workers that your competitors are trying, or will soon try, to develop CI on your firm. Tell them how that can happen. Let them know what your firm's competitively sensitive information is, and remind them that it needs to be protected constantly. Use your own experiences to show them what is vulnerable and how to protect it.

(For additional help in this area, see McGonagle and Vella 1998)

13.3 As For CI's Future—Keep an Eye on Terrorism's Impacts

Just what does terrorism have to do with competitive intelligence, and vice versa? Each has the potential impact in the other in many more ways than we can ever imagine. Let us just list a number of ways that can occur and then develop each of them in turn:

- Businesses are changing and will continue to change the way they deal with outsiders.
- Numerous changes to open records or freedom of information act statutes and regulations, both at the federal and state levels, have been implemented, and will continue to develop.
- People are more sensitive to what other people are doing and are on the lookout for anything "strange".
- There will certainly be other unintended, perhaps even unpredictable, consequences.

13.3.1 Businesses Change the Way They Deal with Outsiders

Changes by businesses in dealing with outsiders have already made it more difficult to collect primary information on many targets. For example, no longer are complete firm directories a regular feature on Internet home pages. No longer are tables of organization and plant diagrams found on the Internet or on the walls in public reception areas in businesses.

Businesses are not doing these things to foil your CI activities. They are doing them to protect against potential terrorist activities. But, in protecting against terrorists, in many cases that also means blocking your CI efforts.

As the war on terror continues, we expect that firms will continue to enhance the protection of employees and the information their employees possess. If and when businesses begin to be the targets of terror, we expect to see such protective and defensive activity escalate to levels that today are almost unimaginable. If such escalation occurs, it will not put competitive intelligence out of business. But what it will do is make the collection of primary intelligence, information from people, or from observations, much more difficult. That in turn means that your CI activities have to rely on secondary or document-based data for more and more of your own intelligence purposes.

While that may not sound like a problem, it is. Basing CI almost exclusively on secondary research is, alas, not unusual today. But the fact that it is done widely does not make it a sound practice. There are many fundamental flaws with this. Among them is that knowledge of what is not yet happened is likely to be in possession of people. Secondary sourcing is almost always relegated to reporting that which has happened in dealing with the future in only the most general terms.

If increasing reliance on secondary sourcing only should become more common in CI, we expect an increase in the use of a particularly corrosive defensive tactic–disinformation. Disinformation is not a topic often talked about in competitive intelligence, in part because of the distaste that even considering it rightly causes in most people.

But disinformation is there, has been there, and will continue to be there. Besides, its obvious ethical problems, which are significant, disinformation has the effect of destroying trust in those spreading it. In business, for disinformation to work, it must mislead competitors, and those persons, or least most of them, spreading it must actually believe in it too. When it becomes obvious later on that the information was incorrect, these people will, consciously or subconsciously, be less willing to do the bidding of those who created the disinformation in the future. In addition, disinformation, if taken too far, can cross over into fraud. But even when it does not cross over, there is a great gray, and, frankly distasteful, area between strict honesty and fraud–disinformation.

13.3.2 There have been Numerous Changes to Open Records Laws

Shortly following the attacks of 9/11/2001, the US government adopted a new, more rigid, policy toward the release of information to the public. For decades, citizens and businesses had come to rely on the US and state freedom of information acts to gain access to government records about themselves and others. Of course, there have always been exceptions to access, such as guaranteeing personal privacy, protecting ongoing criminal investigations, etc. But in general, these laws have been a powerful tool for controlling the behavior of government. It has also been a powerful tool in the hands of CI professionals, using it to dig out details about firms dealing with the government, details about firms in difficulty with the government, or details about firms seeking to influence the government.

Federal agencies immediately began to restrict access to information that might be of assistance to terrorists. For example, the US Environmental Protection Administration had begun to post very detailed information from the numerous reports filed by businesses on the Internet. It also had begun a program of enriching that information by providing details from publicly accessible business directories, and even access to driving direction programs, producing maps of the areas surrounding particular industrial facilities. Following 9/11, that material disappeared from the Internet. The feeling was, correctly, that the data provided too much assistance to potential terrorists in detailing chemicals being used, physical locations and other details.

However, that same data was being used by firms in the chemicals industry to develop accurate and inexpensive profiles on their own competitors. In fact, the utility and power of this data had been such that official representatives of the chemicals industry had actually protested to the EPA before 9/11, seeking to remove that information from places where their competitors could see it. They even engaged a CI firm to show just how powerful and critical that data was to developing competitive intelligence profiles.

But the limitation of access to data by governments will not stop there. As this book is being written, there are moves within the US government to put together working groups to totally reassess the kinds of information that should be available through FOIA and the kinds of information that should be protected. A part of this battle has surfaced in the use of brand new classifications for data not previously seen, marking efforts by governments to protect data, which was not personal, confidential, or subject to a security classification. The result is that the amount of data available through the US FOIA and the speed with which it can be retrieved are both significantly less than a decade ago.

For example, in the recently enacted Dodd–Frank Wall Street Reform and Consumer Protection Act, signed into law in July 2010, there are provisions that shield the US Securities and Exchange Commission (SEC) from certain requests under the FOIA. The changes were caused due to concern that the FOIA would hinder SEC investigations that involved trade secrets of financial companies.

The same is true at the state level. The states, also responding to increasing citizen concern about identity theft and invasions of privacy, have begun efforts to limit or prevent the release of what most people would consider personal data, reversing past steps toward more open records. This has had unintended, but significant, consequences on the availability of data for businesses.

For example, a short time ago, one state amended its open records laws to try and protect personal, sensitive data. Its goal was that records released by state government agencies could not include identification data, such as taxpayer identification numbers, which for people is the same as their social security number, plus telephone numbers, home addresses and similar such information. This all seems to make perfect sense. However, the way the legislation is written, it impacted not only the contents of, but the actual access to, business records.

Let us give you one example. In the insurance industry, applications for approval of new policies, advertisements, and the like, are typically accompanied by a cover letter from the firm filing the documents. That letter will include the name of the individual responsible for the filing, their direct dial phone number and other data. The applications themselves will usually contain some sort of identification number for the firm—often its federal taxpayer identification number, the same number used in its filings with the US Securities and Exchange Commission.

Because of the change in the law, one insurance department found that it was not able to release documents dealing with the filing of new insurance policies, to which the public and competitors had a right of access, without manually redacting telephone numbers, personal names, and taxpayer identification numbers from each of these documents. Since the documents were in Adobe PDF format, that meant withholding the documents completely from the public until the documents could be rescanned with the all of the material data manually blacked out.

Since these forms had been available online as well, that one legal change completely shut down access to all filings of all insurance firms for as far back as the records went. Eventually, that mess was straightened out by a combination of amendments and the application of common sense. But for a period of time, an effort to protect personal identity and personal phone numbers had the effect of shutting down access to literally thousands of pages of previously public information.

But not all changes in recent years have been intended to restrict access to information. For example, the US Securities and Exchange Commission adopted regulation FD, for full disclosure, to force firms to release critical information, all at once to all parties. The firms of course had been required to disclose material information as it is known, but very often that information was disclosed first to a select group of analysts before it was made available to the media and then to the public. The SEC sought to level the playing field by requiring that this information, when released, the released to everyone simultaneously. While the intent was noble, as with so many other moves by governments, there were unintended, rather strange, consequences.

Oddly enough, changes in SEC regulations requiring more public disclosure by publicly they traded businesses, had the reverse impact. While we might initially think that this makes collecting CI easier, it actually has the effect of making those persons who disclose information about publicly traded businesses more sensitive about what they say and to whom. Anyone reading company briefings of analysts, and comparing those briefings with briefings even six years ago will be struck by the extreme caution exercised by every speaker and the extremely general nature of all exhibits and displays, such as PowerPoint presentations.

At the state level, efforts by companies, such as Google, to open up state governments has resulted in more access to state records than in the past. But cost is also an increasing issue at this level.

So, overall, as the war on terrorism continues, we can expect that access to more and more data currently held by governments will be impacted. Most likely, it will be subject to the typical back and forth of politics. That is, access will be increasingly restricted. That will result in a backlash, shifting the trend toward more "openness and transparency". When data that properly should not be released is found to have been released, there will be another shift, and so forth.

13.3.3 People are More Sensitive to What Other People are Doing

At the writing of this book, the US Department of Homeland Security has launched a new campaign: "If You See Something, Say Something™" (US Department of Homeland Security 2011). Let us give you a real example of what that means to CI data collection. Several years ago, a client wanted information about the production capacity of a plant owned by a competitor. The client firm had done a very good background research assessment of what could be found on the Internet and from industry print resources. From that, they found out the nature of the product produced at this plant, the kind of specialized equipment used in the plant, and the plant's ability to shift between and among various types of products. However, what was lacking was a real in-depth assessment of the current size of the plant and the number of pieces of large manufacturing equipment there. These were both critical to profiling the plant's current capacity.

In this case, we decided to go to the township where the plant was located, specifically to the local office that housed zoning and building codes enforcement. Our hope was that the facility's plans for expansion and renovation and the like, which had been filed there, would lead us to be able to assess current conditions and the current manufacturing area in the facility itself.

The plant was located in a community in a rural region. The plant was probably the largest employer in the county, and certainly the largest in the township. When a representative of Helicon went to this office, it was so small that he was given access to the plans, but had to review them at a worktable in the middle of the office. As requested by the township, he signed his name in a public record request log, identified his firm, although that was not required to state law (but was requested by township officials), and proceeded to evaluate the plans.

Another individual on the assignment attempted to call several people within the plant. The background research by the client had helped us to identify names of individuals, such as the plant manager, so an effort was made to elicit additional information on the plant. The plant was obviously busy as every call was referred internally at least twice.

About 24 hours later we received a call from an individual with the parent firm. The parent had found out, as they put it, "about our interest" in the plant and was very upset. We pointed out that, in fact, both individuals had left their names, and clearly identified themselves in all situations. Otherwise the target firm would have had no way of contacting us.

What happened was not so much that the individuals at the plant reported the relatively innocuous calls to headquarters, but rather that someone in the township had a relative or friend working in the plant. They probably wondered why this individual from outside the area was so interested in the plans of the facility and called someone, probably a relative at the plant. That individual, in turn, probably contacted the plant manager or plant security, who, in turn probably called an engineer at headquarters. That in turn resulted in an inquiry made by headquarters of the small number people the plant as to whether or not they had been contacted by Helicon. The answer was, yes. About half of the key people in the plant in fact had been approached by Helicon, but had provided no data of value.

In our experience, the calls that were being made to elicit information from the plant personnel would not normally have triggered an inquiry like this. Rather, it was the curiosity of an individual in the township about the request to see a public record about a major industrial facility that triggered a defensive response. In over 25 years of conducting competitive intelligence research, this is the only case that we can recall where the local government gatekeeper, as it were, actually alerted a target of our interest in public records, our legitimate and proper interest.

We are certain that this will not be the last such event. We absolutely believe that the appropriate caution that has been implanted among Americans and among others around the world, about inquiries from strangers has already had an impact on CI. Realistically, the caution evidenced by Americans and others will not always be at the highest level. It is our opinion that after every actual or threatened terrorist event somewhere in the world, widely reported in the media, this kind of sensitivity will peak, even if for only a short time. Thus we see the anomalous situation that the threat to bomb an airplane flying from London to the United States may temporarily make legitimate collection of some kinds of data for CI in the United States significantly more difficult.

13.3.4 The War on Terrorism will have Unintended and Unpredictable Consequences

One example of this can be seen in a controversy that arose in early 2007. As many competitive intelligence specialists know, Google Earth basically provides satellite photographs of much of the world and even more detailed ground level looks for many urban areas.

January 2007 news reports indicated that Google was talking with British commanders in Iraq after learning that terrorists attacking British bases in Iraq may have been using aerial footage from Google Earth to pinpoint strikes. As a result, Google removed these bases from its service. (The Register Company 2007)

It does not take much imagination to see that Google Earth, a product used by CI analysts, particularly dealing with existing facilities, may be forced to become less useful, less precise, or even inaccessible under certain circumstances and with respect to certain areas.

In this case, the restrictions dealt with areas where United States troops and other military groups were engaged in active conflict. However, it is merely a short step to having Google Earth confronted with a request to tune down its operations due to a suspected terrorist's map targeting not a British regimental headquarters in Iraq, but rather an American oil refinery in Philadelphia.

And at least half of that step has been taken. The AP has reported that some state governments providing maps to Google are now blurred out to protect sensitive facilities. In one of those states, New York,

New York City still refuses to release architectural plans of 2,500 buildings deemed security sensitive. Even the list of buildings [excluded] is not public. (NJ.COM 2011)

References

McGonagle JJ, Vella CM (1998) Protecting your company against competitive intelligence. Quorum Books, Westport

NJ.COM, 22 Aug 2011. Uneven post-9/11 data-cloaking persists (22 Aug 2011), http://www.nj.com/newsflash/index.ssf/story/ap-impact-uneven-post-911-data-cloaking-persists/91f94d29c01f4ef0bc6e24033917b710 Access date 6 Sept 2011

The Register Company. 7 Jan 2007. Google erases British bases in Iraq. theregister.co.uk/2007/01/17/google_erases_brit_bases/Access date 29 Aug 2011

US Department of Homeland SecurityUS Department of Homeland Security. dhs.gov/files/reportincidents/counterterrorism.shtm, Access date 29 Aug 2011

Index